Wolfgang Kox and David Bihari (eds.)

Shock and the Adult Respiratory Distress Syndrome

With 89 Figures

Foreword by Iain Ledingham

Springer-Verlag
London Berlin Heidelberg New York
Paris Tokyo

Wolfgang Kox, MD
Senior Lecturer, Department of Anaesthesia, Charing Cross and
Westminster Medical School;
Honorary Consultant Anaesthetist (Director, Intensive Care Unit),
Charing Cross Hospital, Fulham Palace Road,
London W6 8RF, England

David Bihari, MA, MRCP
Lecturer in Medicine, Department of Medicine, The Middlesex Hospital
Medical School, Mortimer Street, London W1P 7PN, England

Cover illustration: Interstitial oedema in a sheep lung 48 hours after
inhalation injury.

ISBN-13: 978-1-4471-1445-1   e-ISBN-13: 978-1-4471-1443-7
DOI: 10.1007/978-1-4471-1443-7

British Library Cataloguing in Publication Data
Shock and the adult respiratory distress syndrome.
1. Respiratory distress syndrome
I. Kox, W.   II. Bihari, D.
616.2   RC776.R38

Photosetting by Tradeset, Welwyn Garden City, Herts. AL7 1BH

2128/3916-543210

Foreword

The interrelated syndromes of shock and the adult respiratory distress syndrome (ARDS) continue to attract the attention of both clinical and laboratory scientists. This reflects both the size of the problem and its unresponsiveness to current lines of treatment. Doubtless, a greater appreciation of the underlying pathophysiological disturbances during the past two decades has led to appropriate action and increased survival in the early stages but once established these syndromes have remained remarkably immune to a wide spectrum of therapeutic modalities. This observation stresses the importance of prevention but also indicates the need for continued research into the nature of the established syndromes and the means whereby they may be reversed.

Drs Kox and Bihari are to be congratulated on bringing together within the covers of this volume many of the acknowledged European experts in these two fields of investigation. Each author has provided an up-to-date account of his current experimental and clinical research, and their combined contributions makes fascinating reading. Undoubtedly, these are exciting times in the development of understanding of shock and ARDS. Inevitably, more questions are raised than answers provided, but the accumulated knowledge presented here adds significantly to our understanding of this complex biological jigsaw. From this corporate endeavour will come the clinically useful developments of the future and with them the ultimate hope that the term 'refractory' shock may be finally removed from our vocabulary.

Glasgow   Iain Ledingham
September 1987

Preface

There is no doubt that there have been many advances in the care of the critically ill patient, but the management of septic shock and acute respiratory failure is still fraught with problems. There are as many approaches as there are authorities and considerable controversy remains concerning the support and treatment of patients with these conditions. Although the subjects are frequently discussed at endless international symposia, many clinicians come away with the feeling that they have heard only the prevailing conventional dogma!

For these reasons we thought that it might be useful to invite a small number of workers in this field to participate in a workshop in which their latest research and new ideas could be discussed in a critical atmosphere. Such a workshop took place in Cortina d'Ampezzo, Italy, in March 1986, most kindly sponsored by Fisons PLC, UK. Although there were no proceedings of this meeting, it was generally felt that a number of the contributions and parts of the discussion should be brought together, and these form the basis of this book. Some additional chapters have been included in order to cover more aspects in greater depth. Nevertheless, a publication of this nature can never hope to be definitive but reflects the interests of the editors and their contributors.

It is exactly twenty years since Ashbaugh and Petty introduced the term "acute respiratory distress", which has subsequently become the "adult respiratory distress syndrome" (ARDS), to describe an acute lung injury associated with respiratory failure in adults. Since this landmark in description, and the many publications that have followed, some progress has been made in understanding its pathogenesis and the survival of patients has probably improved. However, not only does the interaction of the various mediator systems require to be clearly defined but the diagnosis of the syndrome also remains somewhat imprecise. Moreover, new methods of primary treatment and ventilatory and cardiovascular support have been introduced yet their clinical role needs further clarification in comparable studies.

The term "ARDS" is as invalid in describing acute respiratory failure as is the label "shock" given to many different forms of acute circulatory failure. Although hypoxaemia and hypotension respectively are clinical markers of the two conditions, the lungs and the circulation probably behave very differently according to the primary insult. Indeed, the two are closely linked for lung damage is one of the manifestations of shock. Why this should be so remains obscure, but certainly the lungs are the only organ to receive the entire cardiac output and the first to be perfused by the venous effluent from damaged and infected tissue. Beyond the administration of appropriate antimicrobial agents, the management of septicaemic shock and the ensuing respiratory failure is essentially supportive. Yet there is no consensus concerning the best form of ventilatory or cardiovascular support and these issues are addressed in this book.

Best ventilation and cardiovascular support by themselves can only buy time for the patient during which lung healing can occur. The treatment of an acute lung injury includes not only the inhibition and blockade of the various activated mediator cascades brought about by the primary insult but also the prevention of further iatrogenic lung damage. Naturally, arising from the collective description of acute respiratory failure as "ARDS", has been the concept that one particular drug or one form of ventilation might provide a "miracle" cure for this complex condition. This naive view has been supplanted by the belief that only a "cocktail" of inhibitors in combination with different forms of mechanical support tailored for the individual patient will eventually improve the outcome.

Sadly, as is usual when a subject is studied in depth, few answers are found and more questions arise. But in posing the right questions which can lead on to further investigation, we will perhaps have contributed some insight into the pathogenesis and management of septic shock and ARDS.

London   Wolfgang Kox
1 January 1987   David Bihari

Contents

Section III: Some Aspects of Ventilatory Support

Section IV: Some Aspects of Cardiovascular Support

Section V: The Diagnosis and Prognosis of the Adult Respiratory
Distress Syndrome

Contributors

A. Adam
Department of Clinical Biology, Centre Hospitalier de Sainte-Ode,
6970 Baconfoy-Tenneville, Belgium

L. Avalli, MD
Institute of Anaesthesia, University of Milan,
Via Francesco Sforza 35, 20122 Milan, Italy

D. Bihari, MA, MRCP
Department of Medicine,
The Middlesex Hospital Medical School, Mortimer Street,
London W1P 7PN, UK

L. Bodson, MD
Department of Anaesthesiology and Intensive Care,
University Hospital of Liège, 4000 Liège, Belgium

P. Damas, MD
Department of Anaesthesiology and Intensive Care,
University Hospital of Liège, 4000 Liège, Belgium

G. Deby-Dupont, Lic. Chemistry, Lic. Biochemistry,
Laboratory of Applied Biochemistry,
University Hospital of Liège, 4000 Liège, Belgium

K. J. Falke, MD
Department of Anaesthesiology, University of Düsseldorf,
Moorenstrasse 5, D-4000 Düsseldorf 1,
Federal Republic of Germany

M. E. Faymonville, MD
Department of Anaesthesiology and Intensive Care,
University Hospital of Liège, 4000 Liège, Belgium

H. Forst, MD
Institut für Anästhesiologie, Universität München,
Klinikum Grosshadern, Marchionistrasse 15, D-8000 München 70,
Federal Republic of Germany

P. Franchimont, MD
Laboratory of Radioimmunology,
University Hospital of Liège, 4000 Liège, Belgium

R. Fumagalli, MD
Institute of Anaesthesia, University of Milan,
Via Francesco Sforza 35, 20122 Milan, Italy

J. Gamble, BSc, PhD
Department of Physiology, Charing Cross Hospital,
Fulham Palace Road, London W6 8RF, UK

L. Gattinoni, MD
Institute of Anaesthesia, University of Milan,
Via Francesco Sforza 35, 20122 Milan, Italy

A. Giuffrida, MD
Institute of Anaesthesia, University of Milan,
Via Francesco Sforza 35, 20122 Milan, Italy

A. B. J. Groeneveld, MD
Medical Intensive Care Unit, Free University Hospital,
De Boelelaan 1117, 1081 HV Amsterdam, The Netherlands

W. Kox, MD
Department of Anaesthesia,
Charing Cross and Westminster Medical School;
and Charing Cross Hospital, Fulham Palace Road,
London W6 8RF, UK

U. Kreimeier, MD
Department of Experimental Surgery, Surgical Center
University of Heidelberg, Im Neuenheimer Feld 347,
D-6900 Heidelberg, Federal Republic of Germany

M. Lamy, MD
Department of Anaesthesiology and Intensive Care,
University Hospital of Liège, 4000 Liège, Belgium

A. Lawson, MB, BS
Intensive Therapy Unit, The Middlesex Hospital,
Mortimer Street, London W1P 7PN, UK

J. R. Le Gall, MD
Réanimation Médicale, Hôpital Saint-Louis,
1 Avenue Claude-Vellefaux, 751010 Paris, France

P. P. Lunkenheimer, MD
Department of Experimental Thoraco-vascular and Cardiac Surgery,
University of Münster, Domagstrasse 3, 44 Münster,
Federal Republic of Germany

R. Marcolin, MD
Institute of Anaesthesia, University of Milan,
Via Francesco Sforza 35, 20122 Milan, Italy

C. A. Marshall, MB, BS, MRCP
Anaesthetic Department, Guy's Hospital,
St. Thomas Street, London SE1 9RT, UK

D. Mascheroni, MD
Institute of Anaesthesia, University of Milan,
Via Francesco Sforza 35, 20122 Milan, Italy

C. N. McCollum, MB, ChB, MD, FRCS
Department of Surgery,
Charing Cross and Westminster Medical School;
and Charing Cross Hospital, Fulham Palace Road,
London W6 8RF, UK

K. Messmer, MD
Department of Experimental Surgery,
University of Heidelberg, Im Neuenheimer Feld 347,
D-6900 Heidelberg, Federal Republic of Germany

P. Niederer, Pd.
Institut für Biomedizinische Technik der EHT und Universität Zurich,
Moussonstrasse 18, Zurich, Switzerland

A. Pesenti, MD
Institute of Anaesthesia, University of Milan,
Via Francesco Sforza 35, 20122 Milan, Italy

K. Peter, MD
Institut für Anästhesiologie, Universität München,
Klinikum Grosshadern, Marchionistrasse 15, D-8000 München 70,
Federal Republic of Germany

K. R. Poskitt, MD, FRCS
Department of Surgery, Charing Cross and
Westminster Medical School; and Bristol Royal Infirmary,
Marlborough Street, Bristol BS2 8HW, UK

J. Racenberg, MD
Institut für Anästhesiologie, Universität München,
Klinikum Grosshadern, Marchionistrasse 15, D-8000 München 70,
Federal Republic of Germany

P. Radermacher, MD
Department of Anaesthesiology, University of Düsseldorf,
Moorenstrasse 5, D-4000 Düsseldorf 1, Federal Republic of Germany

H. Redl, PhD
Ludwig Boltzmann Institute for Experimental Traumatology,
Donaueschingenstrasse 13, A-1200 Vienna, Austria

A. Riboni, MD
Institute of Anaesthesia, University of Milan,
Via Francesco Sforza 35, 20122 Milan, Italy

F. Rossi, MD
Institute of Anaesthesia, University of Milan,
Via Francesco Sforza 35, 20122 Milan, Italy

F. Scarani, MD
Institute of Anaesthesia, University of Milan,
Via Francesco Sforza 35, 20122 Milan, Italy

G. Schlag, MD
Ludwig Boltzmann Institute for Experimental Traumatology,
Donaueschingenstrasse 13, A-1200 Vienna, Austria

A. J. Schneider, MD
Department of Internal Medicine, Free University Hospital,
De Boelelaan 1117, 1081 HV Amsterdam, The Netherlands

N. Stroh, Dipl. Ing.
Institut für Grenzflächen- und Bioverfahreustechnik der
Fraunhofergesellschaft, Nobelstrasse 12, 7000 Stuttgart 80,
Federal Republic of Germany

P. M. Suter, MD
Division of Surgical Intensive Care, Hôpital Cantonal Universitaire,
CH-1211 Geneva 4, Switzerland

L. G. Thijs, MD
Medical Intensive Care Unit, Free University Hospital,
De Boelelaan 1117, 1081 HV Amsterdam, The Netherlands

H. Van Aken, MD
Klinik Für Anaesthesie und Operative Intensivmedizin, Universitätsklinik,
Albert Schweitzer Strasse, 44 Münster, Federal Republic of Germany

J. L. Vincent, MD, PhD
Department of Intensive Care, Erasme Hospital,
Free University of Brussels, Route de Lennik 808,
B-1070 Brussels, Belgium

S. Westaby, BSc, MS, FRCS
Department of Cardiac and Thoracic Surgery,
John Radcliffe Hospital, Headington, Oxford OX3 9DU, UK

W. F. Whimster, MD, FRCP, FRCPath
Department of Morbid Anatomy, King's College Hospital
School of Medicine and Dentistry, University of London,
Denmark Hill, London SE5 9RS, UK

Zh. Yang, MD
Department of Surgery, Tongji Hospital, Tongji Medical University,
Wuhan, People's Republic of China; and Department of Experimental
Surgery, University of Heidelberg, Im Neuenheimer Feld 347,
D-6900 Heidelberg, Federal Republic of Germany

Section I

The Morphology and Pathogenesis of Acute Lung Injury

Chapter 1

Fluid Flux Across the Microvascular Endothelium

J. Gamble

Our understanding of the forces governing the movement of fluid into and out of the vascular compartment owes much to the observations of Starling at the end of the nineteenth century (Starling 1894, 1896). The history of the development of the Starling Hypothesis and its subsequent mathematical formulation has been reviewed in successive editions of the *Handbook of Physiology* (Landis and Pappenheimer 1963; Michel 1984).

The object of this chapter is to give a brief survey of the morphological, physicochemical and hydrodynamic principles governing the interaction of the Starling forces across the exchange vessels of the microcirculation. No attempt has been made to review the large body of experimental evidence that is available, for this has been done adequately elsewhere. Where possible review references have been provided to facilitate access to the relevant literature.

The Starling Equation

The transendothelial hydrodynamic state that is called the Starling Balance or Starling Equilibrium is actually a state of dynamic equilibrium for, in tissues that are neither gaining nor losing weight (isogravimetric) the continual net efflux of water and protein from the capillaries into the interstitium is matched by an equal outward movement of fluid and protein from the interstitium via the lymphatics.

The experimental evidence that has been used to explain the Starling forces has been derived from two main sources: (i) whole organ studies and (ii) studies on individual microvessel segments. Each approach has its advantages and disadvantages. Firstly, in both situations the tissues have been subjected to some degree of trauma in order to make the study possible. Secondly, the heterogeneity of microvessels within tissues implies that the results from studies on single microvessel segments may be unrepresentative of the whole, while in the studies on whole organs the lumped values may give a seriously distorted view of the individual segments within that piece of tissue. Despite these obvious disadvantages a considerable amount of information about the relationship between the forces governing the movement of fluid across the endothelium of the microvascular exchange vessels (capillaries and venules) has been obtained and can be summarised by the following equation:

$$J_v/A = L_p[(P_c - P_t) - \sigma(PI_c - PI_t)] \tag{1}$$

where J_v/A = the volume flow per unit area of capillary wall (A),
L_p = hydraulic conductance of the capillary wall, measured in ml min^{-1} per mmHg change in pressure, per unit area of wall,
P_c = capillary hydrostatic pressure,
P_t = interstitial hydrostatic pressure,
PI_c = intravascular colloid osmotic (oncotic) pressure,
PI_t = mean interstitial oncotic pressure,
σ = osmotic reflexion coefficient (a constant defining the selectivity of the endothelium to the solute molecule in question which can be represented as the ratio of measured to predicted oncotic pressure values (see later)).

This equation is used in experiments where single segments of microvessel have been studied under the microscope and the surface area available for fluid flux can be measured quite accurately. In gravimetric or volumetric studies where changes in weight or volume (respectively) of whole organs or pieces of tissue are being studied and the surface area available for exchange is not known, the capillary filtration term used, denoted by K_f, represents the fluid flux (J_v) per unit tissue weight per unit change in filtration pressure. The major limitation of this last technique lies in not knowing how much of the potential microvascular surface area is actually being used.

The significance of the various components of Eq. (1) are probably best understood in relation to the morphology of the exchange vessel–interstitial interface.

The basic requirement for a balance between oncotic and hydrodynamic forces can be satisfied by a membrane acting as a barrier across which oncotic pressures can be developed and also across which the hydrodynamic force, generated by the combination of cardiac work and gravitational forces, is dissipated. For ease of discussion this barrier will be considered under two headings: the primary barrier, which includes the endothelium, its basement membrane and associated structures; and the secondary barrier which includes the interstitium and associated lymphatics.

The Primary Barrier

The Endothelium

The exchange vessels consist of a layer of specialized simple squamous epithelial cells, the endothelium, which is separated from the luminal contents by the glycocalyx and from the interstitium by the basement membrane; the arrangement is shown diagrammatically in Fig. 1.1. The endothelial cells form a thin layer 0.1–0.5 μm thick; the individual cells are elongated polygons, 10–15 μm wide by 25–40 μm long, with the long axis along the length of the vessel. Besides the nucleus, the endothelial

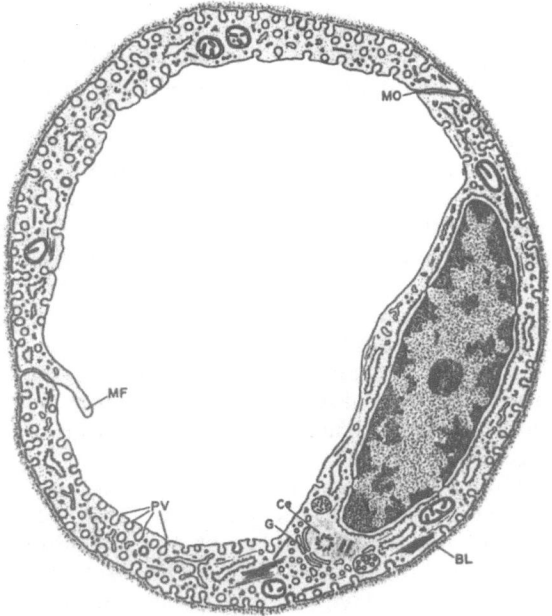

Fig. 1.1. Drawing of capillary cross-section showing the basement membrane (BL), Golgi apparatus (G), intercellular cleft with occluding junction (MO) and marginal fold (MF). The inner luminal surface is coated by glycocalyx (not depicted here: see text). Ce, centrioles; PV, pinocytotic vesicles. (From Lentz (1971), by permission of W. B. Saunders, Philadelphia.)

cells contain a number of other organelles including vesicles and a well-developed rough endoplasmic reticulum which is usually associated with secretory cells and which is believed to increase during the healing phase following injury. Other organelles which are of particular interest because of their possible role during inflammatory reactions are the thin actin filaments, thick myosin filaments and microtubules. It has been proposed that histamine-type mediators cause the endothelial cells, particularly those in venules, to contract (see later).

Glycocalyx

The existence of the glycocalyx or luminal coat has been revealed by staining with ruthenium red, a charged electron-opaque material. The luminal coat is 10–20 nm thick and consists of glycosaminoglycans (secreted by the cell), oligosaccharides, projecting from the cell membrane, together with glycoproteins, glycolipids and sialoconjugates. These surface-lining components are also found in the junctional spaces between the cells. Studies using other charged tracers have shown that the endothelial cell coat has an overall negative charge. There is a non-uniformity of charge distribution across the surface which gives rise to a series of surface microdomains, the functional significance of which is considered by Simionescu and Simionescu (1984) in their recent review.

Basement Membrane

The basement membrane (or basal lamina) of the vessels of the microcirculation is a layer composed largely of collagens, glycoprotein and fibronectin, some of these components being secreted by the endothelial cells; the cells are linked to the basement membrane by fine fibrillar strands. The membrane is 40–80 nm thick and is the interface between the endothelial cells and the interstitium. It forms a continuous layer beneath the endothelium of both fenestrated and continuous capillaries but is discontinuous under the spaces in sinusoidal microvessels. The structure and nature of the basal lamina are considered in detail by Simionescu and Simionescu (1984).

The exchange of materials between the intravascular and interstitial compartments involves penetration of the three components of the primary barrier—namely the glycocalyx, the endothelial cell layer and the basement membrane—before the interstitium is reached. The interstitial distances that have to be traversed before the metabolizing cells are reached can be considerable. The two main mechanisms of solute movement are via diffusive and convective fluxes.

The permeability of the vascular endothelium to water and both small and large hydrophilic molecules varies considerably according to location (see Table 1.1, p. 9). The movement of water outside the endothelial boundary of the circulation can be considered as an "extravascular circulation" which comprises the exchange vessel filtration and subsequent microcirculatory and lymphatic absorption of water and solutes. It is very important for both the maintenance of blood volume and the removal of macromolecular substances from the interstitium via the lymphatics. However, this extravascular circulation is not adequate to cope with the demands for exchange of small solutes between the metabolizing cells and the microcirculation. Metabolic exchange is largely dependent upon diffusive processes and these are almost independent of the magnitude and direction of the net fluid movement.

The movement of water and solutes between the microcirculation and metabolizing cells requires a pressure gradient for the filtration of water and transendothelial pathways with permeabilities sufficient to permit the movement of both solvent and solutes. The major obstruction to the movement of both water and solutes across the three components of the microvessel wall is believed to be the endothelial cell layer itself, and a considerable amount of time has been spent trying to identify specific channels or "pores" which penetrate this layer. Three major potential pathways have been identified and subjected to great scrutiny; they are the intercellular junctions, cellular vesicles and fenestrae, which are depicted diagrammatically in Fig. 1.2.

Intercellular Junctions

The nature of the junction between adjacent endothelial cells is very important, for it has considerable bearing on the permeability of the membrane to water and to micro- and macromolecular solutes. Two main types of junction are described. The first are the occluding (zona occludens) or tight junctions which constitute a physical link between two adjoining cells; these present an occlusive barrier to the intercellular passage of materials. The second are the communicating (gap) junctions (macula communicans). Gaps up to 6 nm wide have been found at these junctions; however, the points of contact are believed to be important for interendothelial cell communi-

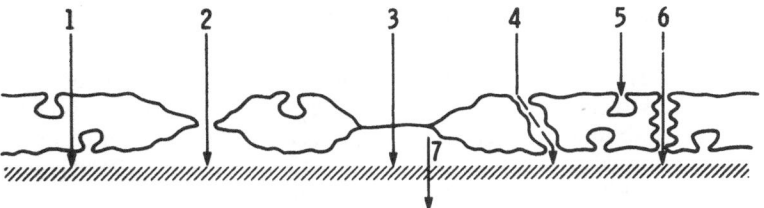

Fig. 1.2. Proposed pathways for the movement of solutes and water across the capillary endothelium: (1) across the whole endothelial cell, (2) through specialized openings in the cell, (3) through endothelial cell fenestrae, (4) through interendothelial cell junctions, (5) by vesicular transport (transcytosis), and (6) via continuous vesicular channels. Having crossed the glycocalyx and endothelium the materials then have to traverse the basement membrane (7). (From Renkin (1977), by permission of the British Medical Association.)

cation and are reported to be involved in intercellular electrotonic and molecular transfer.

Most of the knowledge about the structure of the junctions has been obtained by electron micrographic studies of fixed specimens of tissue. Functional correlation has been obtained by putting electron-opaque tracers of known sizes in the vascular compartment and then, after a period of equilibration, fixing the tissue and making electron micrographic studies of the junctional types and location of the markers. The freeze-fracture technique has also been employed which, as the name implies, involves freezing the tissue, fracturing it and then studying structural components along the lines of cleavage.

These combined structural studies have enabled three-dimensional pictures of the junctional regions to be built up which show them to be far more complex than the two-dimensional picture indicated in Fig. 1.1. It is clear that at best the intercellular pathways for both solute and solvent movement are quite tortuous. The studies have also revealed differences in the cellular junctions in different parts of the microvascular bed. In arterioles there are continuous and elaborate occluding junctions with occasional communicating junctions. In capillaries, branching strands of staggered occluding junctions have been seen but communicating junctions are absent. The junctional pattern changes abruptly on moving from the capillary to the venular segments. The venular junctions appear to vary according to the vascular bed under consideration. Generally the venular junctions seem to be much less tight than the junctions between the endothelial cells in the preceding sections of the microcirculation, 25%–30% of them being classified as communicating junctions. The structural and functional significance of these junctions has been reviewed by Simionescu and Simionescu (1984).

Vesicles

Electron micrographic studies have shown that endothelial cells contain membrane-lined spherical inclusions of uniform diameter (60–80 nm). These structures have been found to be either free in the cytoplasm or opening onto either the luminal or abluminal surface. Whilst plasmalemmal vesicles are present in all endothelial cells there are characteristic differences in population density between vascular beds; the numbers also appear to increase progressively from the arterial to the venular end of the microvascular beds.

The function of vesicles has been the subject of vigorous debate. Two main functions have been ascribed to them: increase in available cell surface area for interaction between membrane-bound molecules (enzymes/receptors) and solutes in the plasma; and translocation of substances across the endothelial cell. Three major theories have been proposed to account for the latter function:

1. *The "ferry boat" theory.* According to this theory vesicles form and are filled on one side of the cell and then move to the opposite side of the cell to discharge their contents.

2. *The "fission–fusion" theory.* This theory proposes that the vesicles are relatively static but are capable of fusing with either the outer membrane of the cell (thereby facilitating emptying or filling) or with adjacent cytoplasmic vesicles, thus facilitating the diffusive movement of the vesicular contents across the cell.

3. *The "continuous channel" theory.* This theory proposes that some of the vesicles shown in electron micrographs are transections through rather tortuous, but continuous transendothelial channels.

There is considerable support for the idea that vesicles are static, rather than mobile structures (Frokjaer-Jensen 1980), so that the ferry boat theory has little support. Transendothelial channels giving the appearance of two centrally fused vesicles have been observed at the venular end of capillaries by some workers; these and other structural aspects of endothelial permeability are reviewed by Palade et al. (1979) and more recently by Simionescu and Simionescu (1984). Labelling studies have shown that vesicles are implicated in the transfer of ferritin (Bruns and Palade 1968), dextrans and glycogens, with molecular sizes ranging from 10 to 30 nm (Simionescu et al. 1972), from the luminal to the abluminal side of the capillary. Bundgaard et al. (1979) showed labelling of free cytoplasmic vesicles in fixed tissues; subsequent electron micrographic study of serial sections of such material (Frokjaer-Jensen 1980) showed that in capillaries studied, whilst 70% of vesicles appeared not to be in contact with the surface in single section studies, more than 99% had indirect contact with the surface when assessed by serial sectioning. In later studies Bundgaard and Frokjaer-Jensen (1982) showed that whilst many fused vesicles were present, they only rarely formed continuous channels between the luminal and abluminal surfaces. The evidence offered by these authors seriously challenges the hypothesis that vesicles have any significant role at all in the transendothelial transport of small solutes.

Fenestrae

Fenestrae are discrete depressions in the opposing luminal and abluminal surfaces of the endothelial cell. The endothelial cell thickness of fenestrated capillaries is less than that of non-fenestrated capillaries and the final thickness of the endothelial cell across the fenestral diaphragm may be only 5 nm. Fenestrated capillaries are mostly found in tissues with high water permeability requirements (e.g. kidney, exocrine glands, synovium, choroid plexus and intestinal mucosa), but they have also been found in tissues where the volume flux is presumably usually low (e.g. endocrine glands). Tissue capillaries with lower water permeability requirements, such as those of lung, skeletal muscle, skin and brain, have few or no fenestrae. Fenestrae lack the lipid bilayer over the surface though there is usually evidence of an electron-dense layer, or diaphragm, over the fenestrae. Charge studies indicate that this region is highly negatively charged. Casley-Smith (1983) has suggested that the structures might be transitory, i.e. continually deforming and re-forming.

A recent analysis of the permeability properties of fenestrae (Levick and Smaje 1986) found that the properties of the diaphragm did not account adequately for the calculated permeability of the fenestrae, implying that other structures, such as the basement membrane and glycocalyx, also played a part in determining the overall permeability characteristics at these sites.

In this section we have seen that there are a number of different types of channel, that run in parallel across the endothelial membrane; differences in their population density confer differences in the measured permeability. It is suggested that some of the channels are permeable to water and small ions whilst the others, possibly the transcellular vesicle pathways, also enable the diffusive flux of macromolecules. Present knowledge about the permeability of the microvascular endothelium is probably adequate to account for the movement of small solutes and albumin, but not adequate to account for the permeability to larger macromolecules such as the immunoglobulins; the search for a small number of larger pores that would permit the movement of these larger molecules continues.

The hydraulic conductance of the microvascular membrane can be represented by the filtration coefficient (K_f). Examples of values of K_f found in different tissues are presented in Table 1.1. The table also contains values for the osmotic reflexion coefficients for albumin (σ). It can be seen that the osmotic reflexion coefficient for albumin across fenestrated endothelium is similar to that seen in non-fenestrated capillaries ($\sigma = 0.8-0.9$), despite very large differences in K_f. Since there are considerable differences in the nature of the interendothelial junctions and the distribution

Table 1.1. Values for filtration coefficients, surface area and sigma in capillaries from different tissues

	K_f (ml min^{-1} mmHg^{-1} 100 g^{-1})	Surface area (cm^2 g^{-1})	L_p (cm s^{-1} dyne^{-1} ×10^{10})	σ
Salivary gland	1.3	512	3.1	0.8
Intestinal mucosa	0.7	276	3.2	0.9
Cardiac muscle	0.34	500	0.86	0.8
Skeletal muscle	0.04	90	0.55	0.91

The values of filtration coefficient (K_f) of the fenestrated capillaries of salivary gland and intestinal mucosa compared with those of the non-fenestrated capillaries of cardiac and skeletal muscle show that even when the values are corrected for surface area to give the value for hydraulic conductivity (L_p), there is still a large difference between fenestrated and non-fenestrated vessels. Despite differences in hydraulic conductivity the values for σ remain similar.

of fenestrae and vesicles, these observations suggest that some structural component of the microvessel walls might be giving rise to the observed conformity in sieving properties. The two structures that are common to all of these vessels are the glycocalyx and the basement membrane. Curry and Michel (1980) have suggested that the structural component that gives rise to these properties is the plasma membrane or glycocalyx which, by acting as a "sieve", restricts the outward movement of macromolecules. Restriction depends not only on the size of the macromolecule, but also on the charge (Clough and Michel 1981; Turner et al. 1983).

The Secondary Barrier

In the preceding section we saw that the primary barrier in the exchange vessels of the microcirculation comprises at least three components in series: an endocapillary layer or glycocalyx which is heterogeneous and charged; a central endothelial layer which contains both pre-formed channels (intercellular junctions) and other channels (vesicles, fenestrae and continuous channels); and a sub-endothelial layer or basement membrane, which is also a heterogeneous, charged structure.

Material that can penetrate these barriers will only continue to do so if concentration gradients (for diffusive fluxes) and hydrostatic gradients (for convective fluxes) are maintained between the intraluminal and interstitial compartments. The interstitium must, therefore, be permeable to these substances and cleared of them by either metabolic usage or by microvascular reabsorption (for small molecular weight solutes) or lymphatic drainage (for larger solutes).

The Interstitium

The interstitium is a complex structure; whilst it may vary in detail from one tissue to another, the basic structural components—collagen and elastin fibres in a ground substance composed largely of hyaluronate, chondroitin sulphate and proteoglycans (Bert and Pearce 1984)—are probably common to all tissues. The structural molecules of the ground substance give rise to a colloid osmotic pressure in their own right. Furthermore, the structural arrangement of the hyaluronic acid molecules, which are long, thin and multiply coiled, give rise to a two-phase, or two-compartment system.

The phase containing the hyaluronate coils, which is freely permeable to water and ions, will, because of the molecular spacing and charge, exclude the migrating macromolecules so that the protein distribution will be limited to one compartment only, thereby elevating the effective colloid concentration within that compartment. About 58% of the body albumin is found in the extravascular space (Rothschild and Waldmann 1970), so that the interstitium can be considered as a large extravascular reservoir of both water and colloidal components. Moreover, due to exclusion, only about 50% of the interstitial space is available for albumin (see later).

Diffusional fluxes of materials result from random kinetic motion of molecules down concentration gradients. In the interstitium the presence of fibrillar components and other interstitial and solute molecules will impede molecular flux by frictional hindrance and by steric hindrance, the significance of these effects being determined by the size, shape and charge of both solute and impeding molecules.

The flow and distribution of fluid within the interstitial space is not well understood. It is generally accepted that the interstitium is a multi-compartment system, but it is not known whether the compartments run in parallel or in series with respect to the direction of fluid and macromolecular flux. It has been suggested that there is a random network of channels within the interstitium which enables water and proteins to move towards the lymphatic channels. The evidence for the existence and function of these channels has been reviewed by Casley-Smith (1983).

The methods of measuring interstitial pressures and the significance of the values obtained has been the subject of much controversy, mostly because the methods used to make the studies produce morphological disturbance in the immediate environment of the intended study. The topic has been reviewed by Brace (1981). Values for interstitial pressure range from the sub-atmospheric to above atmospheric. Whatever value is achieved, development of oedema results in an elevation of measured pressure.

For interstitial pressures to remain sub-atmospheric implies that fluid is continually being removed from the interstitial space (Zweifach and Lipowski 1984). The two main reabsorptive pathways are via exchange vessels, using the Starling forces, which removes water and small solutes; and via the lymphatics, which remove macromolecular components together with water. There is no evidence to support the capillary reabsorption of macromolecules.

The precise mechanisms of lymph formation and the factors governing it have yet to be evaluated. It is believed that the movement of fluid into the lymphatic capillaries is dependent upon relatively small shifts in interstitial hydraulic and oncotic pressure. Clough and Smaje (1978) recorded negative pressures in the terminal lymphatics of cat mesentery when the surface was superfused with osmotically inert fluorocarbon, whereas when this was replaced by Krebs–Ringer solution, which was imbibed by the tissue, the pressures became marginally positive.

Lymphatic Pumps

Micropuncture studies of the pressures in terminal lymphatics have given control pressures ranging from −1.0 to +0.5 mmHg. As one proceeds from the terminal to the major lymphatic vessels the recorded pressures increase as the luminal diameter increases. These observations show that the pressure gradient is the reverse of that required to maintain flow. The larger lymphatic vessels of a number of species have been shown to contain smooth muscle cells and to undergo rhythmic contractions. They have also been observed to respond myogenically, so that increases in intraluminal pressure, which tend to stretch the vessel, stimulate contractions. The lymphatic vessels contain valves, so that increases in intraluminal pressure force the lymph onwards towards the systemic circulation. Periodic contraction or pulsation of associated structures (e.g. skeletal muscle or adjacent arteries) assists in the return of lymph by increasing the transmural pressure. Sustained increases in interstitial pressure tend to compress the lymphatics and increase the resistance to the return flow of lymph.

Cannulation of lymphatic vessels and the collection of lymph has provided a lot of information about the selective permeability of the series barriers between the intra-

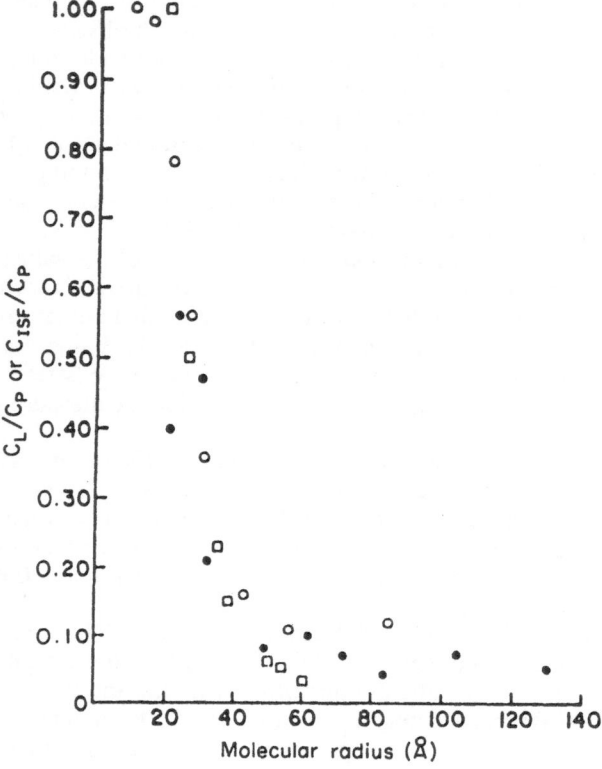

Fig. 1.3. The relationship between the molecular radius of test solutes and their steady-state distribution between the intravascular and interstitial compartments. C_L, concentration in lymph; C_P, concentration in plasma; C_{ISF}, concentration in interstitial fluid. *Filled circles and squares* are data from dog paw lymph; *open circles* are data from rabbit interstitial fluid. (From Guyton et al. (1975), by permission of W. B. Saunders, Philadelphia.)

vascular and lymphatic compartments and also about the transit times for the movement of macromolecules between the intravascular compartment and the lymph. This information has been reviewed by Taylor and Granger (1984). In essence it has been demonstrated that the movement of macromolecules between intravascular and lymphatic compartments is dependent upon the size of the macromolecules under consideration. These results are well illustrated in Fig. 1.3. It should be emphasized that these data give no information about the relative permeabilities of the series barriers between the microvascular and lymphatic channels.

Compliance

Studies by Guyton (1965) have demonstrated a sigmoid relationship between tissue pressure and volume. At sub-atmospheric pressures the interstitial compliance is low and small changes in volume are accompanied by large changes in pressure. When the interstitial pressure becomes atmospheric, the compliance appears to change and

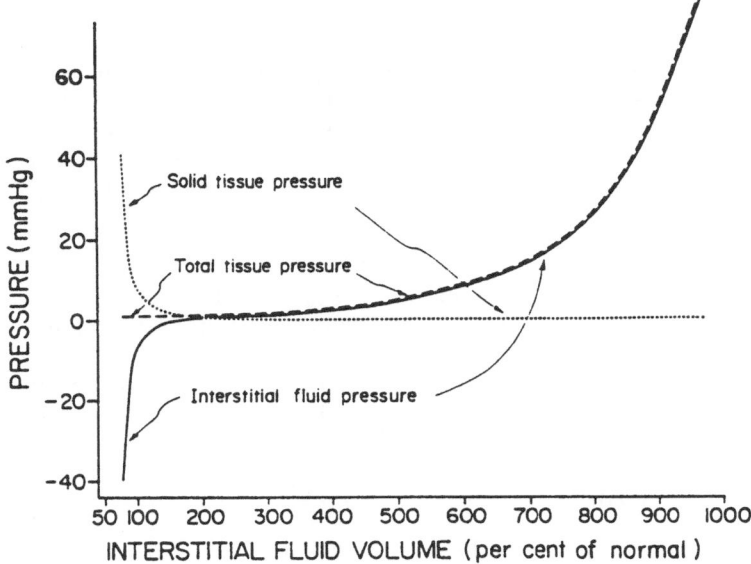

Fig. 1.4. The relationship between interstitial fluid volume and the pressures in the solid tissue, total tissue and interstitial fluid compartments within the tissue. (From Guyton et al. (1975), by permission of W. B. Saunders, Philadelphia.)

small changes in pressure are accompanied by large changes in tissue volume. As the tissue volume increases, the compliance appears to decrease gradually, so that changes in volume are accompanied by larger increases in pressure; the relationships are illustrated in Fig. 1.4. Guyton's observations were made in subcutaneous tissue. A similar relationship has been observed in the interstitium of synovial joints by Knight and Levick (1983).

One possible explanation for the observed relationship is that the first apparent compliance change denotes an alteration in interstitial interfibre bonding allowing areas that were formerly relatively fluid-free to expand and become hydrated. The final phase probably marks the limitations of an outer bounding tissue lining or capsule.

Fluid Flux

Having outlined the major morphological components of the "primary" and "secondary" microcirculatory barriers, I will now consider the hydrodynamic and physicochemical basis for the Starling balance of forces across them. It can be deduced from Eq. (1) that any alteration in the transendothelial oncotic or hydrostatic forces will result in the movement of fluid which will continue until a new state of dynamic equilibrium is achieved. Alteration of the capillary filtration coefficient will only alter the time taken to reach this equilibrium.

Colloid Osmotic Forces

The combination of a microvascular lining which is relatively impermeable to proteins and a high intravascular protein concentration gives rise to an intravascular colloid osmotic (oncotic) pressure which results in a net force drawing fluid into the vascular lumen from the surrounding interstitium. The relationship between macro-molecular concentration and oncotic pressure is complex; for small molecular weight ideal solutes in low concentrations it is given by the van't Hoff equation:

$$\text{Osmotic pressure} = cnRT \qquad (2)$$

where osmotic pressure is measured in atmospheres, c is the molar concentration of solute, n the number of particles produced per molecule by ionization, R the univer-sal gas constant and T the absolute temperature.

Larger molecular weight species are non-ideal and the relationship between concentration of solute and the pressure generated is non-linear. Landis and Pappenheimer (1963) showed that the relationship between bovine serum albumin concentration and colloid osmotic pressure (COP) could be described by the equation

$$\text{COP} = 2.8c + 0.18c^2 + 0.012c^3 \qquad (3)$$

where the units for COP are mmHg and c is the concentration in g 100 ml^{-1}. The relationship is made more complex because of the effect of the charge of the protein molecule and the interaction between the protein molecules themselves, on the one hand, and the interaction between ions in solution and the protein molecules on the other. The result of these interacting forces is that the pressure generated by a par-ticular protein concentration will depend upon the temperature, ionic composition and pH of the solution being studied.

Protein molecules are complex charged structures the molecular configuration, and therefore net surface charge, of which depends upon the ionic composition and pH of the solution in which they are suspended. Different proteins have different charge characteristics so that the relationship between concentration and oncotic pressure will differ for each molecular species. Complex mixtures of proteins, such as plasma, will have concentration/oncotic pressure relationships that are dependent upon the ratios of molecular species as well as the ionic composition and pH of the solution.

The Osmotic Reflexion Coefficient, σ

In vitro measurements of oncotic pressure can be made using membrane osmometers (e.g. Hansen 1961). The measurements are best made under ideal conditions of tem-perature and pH and using an osmometer membrane which is known to be imper-meable to all of the macromolecular species in the solution (i.e. its osmotic reflexion coefficient is equal to unity). Thus it can be seen that Eq. (2) can be rewritten:

$$\text{Osmotic pressure} = \sigma\,(cnRT) \qquad (4)$$

where σ is a measure of the ability of the solute to permeate the membrane. If σ is zero, the membrane pores are very large and do not impede the movement of the solute molecules through the membrane so that no osmotic pressure is generated; if σ is 1.0 then the movement of solute molecules is totally restricted and the osmotic pressure will equal the value that would be predicted on the basis of van't Hoff's Law.

The sieving properties of membranes depend upon their structural and physico-chemical characteristics. These features are relatively easy to measure in the artificial membranes used in membrane osmometers; some of the membranes have pores of uniform size whereas others are less specific, having a range of pore sizes. On the basis of this information, Eq. (3) can be rewritten to include σ:

$$COP = \sigma (2.8c + 0.18c^2 + 0.012c^3) \qquad (5)$$

From Eq. (1) it can be seen that in tissues that are neither gaining nor losing weight the effective osmotic pressure in vivo will equal the capillary pressure (Pappenheimer and Sotto-Rivera 1948). The ratio of the value thus obtained to the value obtained using a membrane osmometer, equipped with a membrane which is known to be impermeable to all of the macromolecules in the solution, will give the value of the reflexion coefficient for the microvascular membrane under consideration.

In vivo charge-based interactions between the fibrous components of a membrane matrix or between the components of the matrix and solute molecules may have a significant influence upon the permeability characteristics of the membrane (e.g. Michel and Phillips 1985). Moreover, the presence of a membrane charge will alter the partitioning of smaller molecular species which, because of the Donnan effect, will give rise to an increase in the effective osmotic pressure generated across the membrane (Curry 1984).

Since charge and size of both membrane pore and solute molecule are important in determining the reflexion coefficient, it will be apparent that a given value of reflexion coefficient is, by definition, specific to particular molecular species and the physicochemical conditions under which it was measured.

Hydrostatic Pressure

The oncotic pressure of the plasma proteins gives rise to a net inward force, which tends to draw fluid into the capillary. This force is opposed by a net outward force, the capillary hydrostatic pressure.

The capillary hydrostatic pressure results from the sum of the forces generated by cardiac contraction and the hydrostatic pressure due to the vertical distance between the capillary and the heart, the latter component being applied at both arterial and venous ends of the capillary. The effective capillary pressure depends upon the total pressure drop across the vasculature and the ratio of pre- to post-capillary resistances (Pappenheimer and Sotto-Rivera 1984).

Variations in Capillary Filtration Coefficient and σ

Under normal physiological circumstances the intravascular and extravascular hydrostatic and oncotic forces are the only variables. Under pathophysiological cir-cumstances however, both capillary filtration coefficient and σ may change. Majno et al. (1969) observed that endothelial cells contracted in response to the application of histamine, serotonin and bradykinin. It is suggested that the contraction is due to the stimulation of the actinomyosin-like fibrils which are known to exist in endothelial cells (Becker and Murphy 1969). Arturson (1979) showed that there was an increase in microvascular permeability to both water and macromolecules in response to thermal injury. Swayne and Smaje (1986) observed that histamine, in low concentrations, caused an increase in the value of K_f without altering σ whereas at higher concentrations σ also changed.

These observations can be interpreted by suggesting that at the lower concentration the increase in the width of the interendothelial cell gap, caused by activation of the intracellular contractile components, was not enough to disrupt the glycocalyx, so that whilst the hydraulic conductivity increased, σ remained the same. At higher concentrations of histamine the glycocalyx was disrupted and so both parameters changed. These contractile components may also be implicated in the increased permeability seen in endotoxin-induced shock. Further details of the proposed mechanisms can be seen in the review by Movat (1984).

Capillary Filtration Coefficient (K_f) and pH

It has been observed (Gamble 1983) that a reduction in pH from 7.4 to 7.05 more than doubled the value of K_f that was obtained in isolated perfused rat mesenteries. Values of systemic arterial pH of 7.25 and less are not infrequently encountered in diabetic ketoacidosis; under these circumstances pH values of 7.05 may well be seen at the capillary interface of hypoperfused tissues.

Protein Effect

Protein itself appears to be important for the maintenance of a normal endothelial filtration coefficient. Studies on mammalian tissues have shown that when perfusing with non-protein colloids the addition of more than 0.2 g% of either plasma protein colloids or bovine serum albumin (concentrations that will be insignificant additions to the total colloid osmotic pressure of the solutions) halves the filtration coefficient (Kinter and Pappenheimer, cited by Landis and Pappenheimer 1963; Rippe and Folkow 1977; Gamble 1978; Diana et al. 1980; Watson 1981). It has been suggested that protein might act by binding with the molecules of the glycocalyx, thereby ordering the glycoprotein chains of the surface coat into a regular array (Michel and Phillips 1985).

An increase in filtration coefficient will result in an increase in the rate of extravasation of fluid in response to an elevation of either the transvascular hydrostatic pressure or oncotic pressure gradients. Fluid efflux will give rise to an elevation of interstitial hydrostatic pressure and a fall in interstitial oncotic pressure. These changes will be accompanied by progressive oedema and an increased lymph flow from the tissue concerned. Tissues differ in their capacity to increase lymph formation and as soon as this capacity is exceeded the fluid accumulation results in oedema.

It can be seen from Eq. (1) that an increase in the rate of fluid filtration (J_v) can be brought about by appropriate changes in either the hydrostatic pressure gradient or effective oncotic pressure gradient across the microvascular wall. The dynamic nature of these changes is highlighted in the paper by Hinghofer-Szalkay and Moser (1986), who observed as much as a 14% decrease in plasma volume and concomitant increase in haematocrit in human subjects during 45 minutes of head-up tilt on a tilt table; the change in volume distribution is attributed to increases in capillary hydrostatic pressure in the dependent capillaries. Eq. (1) shows that a compensatory increase in interstitial pressure should prevent this fluid efflux. Experiments along these lines are considered in a review by Epstein (1978). Subjects moved from the horizontal to the vertical position, but in the latter were immersed up to the neck in water, which thereby produced appropriate counteracting hydrostatic forces in the interstitium at each vertical level in the body. The results obtained in these experiments were far from uniform, but the trend appeared to be towards a decrease, rather than an increase, in haematocrit, suggesting a movement of fluid into rather than out of the vasculature.

A further example of an in-vivo mechanism for preventing oedema—and its failure—is highlighted in studies on fluid filtration in the feet of diabetic and normal subjects (Rayman et al. 1987). In these studies foot dependency gave rise to an increased fluid filtration which was evidenced by foot swelling and an increased colloid osmotic pressure of the blood in the veins leaving the foot. In control subjects the swelling rate was less than would have been predicted on the basis of the increase in capillary pressure. This discrepancy was attributed to the marked decrease in blood flow that was observed, for at low flow rates fluid filtration gave rise to an increase in colloid osmotic pressure that was adequate to counteract the elevated hydrostatic pressure; furthermore the fluid filtration rate was small enough to be dealt with by the lymphatics without excessive foot swelling. In diabetics, however, where neuropathy had diminished the reflex pre- and post-capillary resistance changes, the flow reduction in response to dependency was severely impaired. In these subjects the measured increase in colloid osmotic pressure was not so marked, so that fluid filtration exceeded the lymphatic drainage ability and a predictably greater rate of foot swelling was observed.

Conclusion

In this chapter the morphological, hydrodynamic and physicochemical characteristics that feature in the establishment of an equilibrium between the hydrostatic and colloid osmotic forces across the microvascular interface have been outlined. The dynamic nature of this relationship has been stressed, as has the non-uniformity of the morphology of the different microvascular beds in the body. The development of oedema has been related to the compliance and lymphatic drainage of the tissues under consideration.

References

Arturson G (1979) Microvascular permeability to macromolecules in thermal injury. Acta Physiol Scand [Suppl] 463:111–122

Becker CG, Murphy GE (1969) Demonstration of contractile protein in endothelium and cells of heart valves, endocardium, intimal arteriosclerotic plaques and Aschoff bodies of rheumatoid heart disease. Am J Pathol 55:1–37

Bert JL, Pearce RH (1984) The interstitium and microvascular exchange. In: Renkin EM, Michel CC (eds) Handbook of physiology, sect 2, Cardiovascular system, vol IV, Microcirculation part 1. Waverley Press, Baltimore, Maryland, pp 521–547

Brace RA (1981) Progress towards resolving the controversy of positive versus negative interstitial fluid pressure. Circ Res 49:281–297

Bruns RR, Palade GE (1968) Studies on blood capillaries. II. Transport of ferritin molecules across the wall of muscle capillaries. J Cell Biol 37:277–299

Bundgaard M, Frokjaer-Jensen J (1982) Aspects of the ultrastructure of terminal blood vessels: a quantitative study of consecutive segments of the frog mesenteric microvasculature. Microvasc Res 23:1–30

Bundgaard M, Frokjaer-Jensen J, Crone C (1979) Endothelial plasmalemmal vesicles as elements in a system of branching invaginations from the cell surface. Proc Natl Acad Sci USA 76:6439–6442

Casley-Smith JR (1983) The structure and function of blood vessels, interstitial tissues and lymphatics. In: Foldi M, Casley-Smith JR (eds) Lymphangiology. Schattauer Verlag, Stuttgart New York, pp 27–164

Clough GE, Michel CC (1981) The role of vesicles in the transport of ferritin through frog endothelium. J Physiol (Lond) 315:127–142

Clough GE, Smaje LH (1978) Simultaneous measurement of pressure in the interstitium and terminal lymphatics of the cat mesentery. J Physiol (Lond) 283:457–468

Curry FE (1984) Mechanics and thermodynamics of transcapillary exchange. In: Renkin EM, Michel CC (eds) Handbook of physiology, sect 2, Cardiovascular system, vol IV, Microcirculation part 1. Waverley Press, Baltimore, Maryland, pp 309–374

Curry FE, Michel CC (1980) A fibre matrix model of capillary permeability. Microvasc Res 20:96–99

Diana JN, Keith BJ, Fleming BP (1980) Influence of macromolecules on capillary filtration coefficients in isolated dog hindlimbs. Microvasc Res 20:106–107

Epstein M (1978) Renal effects of head-out water immersion in man: implications for an understanding of volume homeostasis. Physiol Rev 58:529–581

Frokjaer-Jensen J (1980) Three-dimensional organization of plasmalemmal vesicles in endothelial cells. An analysis by serial sectioning of frog mesenteric capillaries. J Ultrastruct Res 73:9–20

Gamble J (1978) The effects of bovine serum albumin on the vascular permeability of the perfused rat mesentery. J Physiol (Lond) 285:15–16P

Gamble J (1983) Influence of pH on capillary filtration coefficient of rat mesenteries perfused with solutions containing albumin. J Physiol (Lond) 339:505–518

Guyton AC (1965) Interstitial fluid pressure. II. Pressure–volume curves of interstitial space. Circ Res 16:452–460

Guyton AC, Taylor AE, Granger HJ (1975) Circulatory physiology, vol II, Dynamics and control of the body fluids. WB Saunders, Philadelphia London Toronto

Hansen AT (1961) A self recording electronic osmometer for quick measurements of colloid osmotic pressure in small samples. Acta Physiol Scand 53:197–213

Hinghofer-Szalkay H, Moser M (1986) Fluid and protein shifts after postural changes in man. Am J Physiol 250:H68–75

Knight AD, Levick JR (1983) Time-dependence of the pressure–volume relationship in the synovial cavity of the rabbit knee. J Physiol (Lond) 335:139–152

Landis EM, Pappenheimer JR (1963) Exchange of substances through the capillary walls. In: Hamilton WF, Dow P (eds) Handbook of physiology, sect 2, Cardiovascular system, vol II. American Physiological Society, Washington, DC, pp 961–1034

Lentz TL (1971) Cell fine structure. An atlas of drawings of whole-cell structure. WB Saunders, Philadelphia London Toronto

Levick JR, Smaje LH (1986) An analysis of the permeability of a fenestra. Microvasc Res 33:233–256

Majno G, Shea SM, Leventhal M (1969) Endothelial contraction induced by histamine-type mediators. J Cell Biol 42:647–672

Michel CC (1984) Fluid movements through capillary walls. In: Renkin EM, Michel CC (eds), Handbook of physiology, sect 2, Cardiovascular system, vol IV, Microcirculation part 1. Waverley Press, Baltimore, Maryland, pp 375–409

Michel CC, Phillips ME (1985) The effects of bovine serum albumin and a form of cationised ferritin upon the molecular selectivity of the walls of single frog capillaries. Microvasc Res 29:190–203

Movat HZ (1984) Microcirculation in disseminated intravascular coagulation induced by endotoxins. In: Renkin EM, Michel CC (eds) Handbook of physiology, sect 2, Cardiovascular system, vol IV, Microcirculation part 1. Waverley Press, Baltimore, Maryland, pp 1047–1076

Palade GE, Simionescu M, Simionescu N (1979) Structural aspects of the permeability of the microvascular endothelium. Acta Physiol Scand [Suppl] 463:11–32

Pappenheimer JR, Sotto-Rivera A (1948) Effective osmotic pressure of the plasma proteins and other quantities associated with the capillary circulation in the hind-limb of cats and dogs. Am J Physiol 152:471–491

Rayman G, Williams SA, Gamble J, Tooke JE (1987) A study of the factors governing fluid filtration in the feet of diabetic and normal subjects. Int J Microcirc Clin Exp (submitted for publication)

Renkin EM (1977) Multiple pathways of capillary permeability. Circ Res 41:735–743

Rippe B, Folkow B (1977) Capillary permeability to albumin in normotensive and spontaneously hypertensive rats. Acta Physiol Scand 101:72–83

Rothschild MA, Waldmann T (eds) (1970) Plasma protein metabolism. Regulation of synthesis, distribution and degradation. Academic Press, New York

Simionescu M, Simionescu N (1984) Ultrastructure of the microvascular wall: functional correlations. In: Renkin EM, Michel CC (eds) Handbook of physiology, sect 2, Cardiovascular system, vol IV, Microcirculation part 1. Waverley Press, Baltimore, Maryland, pp 41–102

Simionescu N, Simionescu M, Palade GE (1972) Permeability of intestinal capillaries. Pathway followed by dextrans and glycogen. J Cell Biol 53:365–392

Starling EH (1894) The influence of mechanical factors on lymph production. J Physiol (Lond) 10:14–155

Starling EH (1896) On the absorption of fluid from connective tissue spaces. J Physiol (Lond) 19:312–326

Swayne GTG, Smaje LH (1986) Histamine can increase filtration coefficient without changing reflexion coefficient to albumin in single rat venules. Int J Microcirc Clin Exp 5:198

Taylor AE, Granger DN (1984) Exchange of macromolecules across the microcirculation. In: Renkin EM, Michel CC (eds) Handbook of physiology, sect 2, Cardiovascular system, vol IV, Microcirculation part 1. Waverley Press, Baltimore, Maryland, pp 467–520

Turner MR, Clough GE, Michel CC (1983) The effects of cationised ferritin upon filtration coefficients of single frog capillaries. Evidence that proteins in the endothelial cell coat influence permeability. Microvasc Res 25:205–222

Watson PD (1981) The effect of protein and dextran on capillary filtration coefficient in isolated cat hindlimb. Microvasc Res 21:261–262

Zweifach BW, Lipowski HH (1984) Pressure–flow relations in blood and lymph microcirculation. In: Renkin EM, Michel CC (eds) Handbook of physiology, sect 2, Cardiovascular system, vol IV, Microcirculation part 1. Waverley Press, Baltimore, Maryland, pp 251–308

Chapter 2

The Morphology of the Adult Respiratory Distress Syndrome

G. Schlag and H. Redl

The morphology of adult respiratory distress syndrome (ARDS) essentially involves two distinct aspects: aetiology and pathophysiology. Many different pathophysiological mechanisms may be important in the development of ARDS. However, although during the initial phase the morphological appearances may be somewhat disparate, eventually the morphological picture is that of classical ARDS.

The primary changes are of great relevance, since they represent a frequent, often reversible process and, if correctly treated at the appropriate time, organ function can be fully restored. The most important part of the lung in morphological terms is the alveolar septum, where gas exchange takes place. This is also where early pathological changes occur, while extension to the interstitial spaces is observed as a later event. Light microscopy is not always able to demonstrate this change in cellular structure and the initiation of cellular damage, and only the advent of electron microscopy has resolved some of these problems.

The alveolar septa divide the alveolar region into individual alveoli. The alveolar septum consists of: epithelium (type I and type II pneumocytes), interstitium, basal membrane and the endothelium of the alveolar capillary vessels (Fig. 2.1).

Pathogenesis of Post-traumatic ARDS: Direct Versus Indirect Lung Injury

In the pathogenesis of ARDS a distinction is made between direct and indirect lung injury. Injury may also occur from a combination of both these processes. According to its aetiology the pulmonary damage differs primarily in morphology.

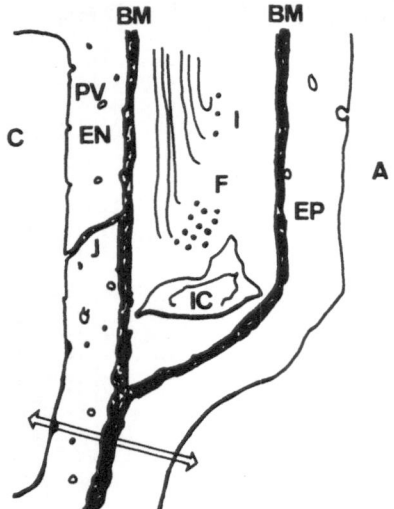

Fig. 2.1. Schematic diagram of the blood–gas barrier in the alveolar septum. A, alveoli; C, capillary lumen; I, interstitium with interstitial cell (IC) and collagen and elastic fibres (F); BM, basal membrane; EN, endothelial cells with cell junction (J) and pinocytotic vesicles (PV); EP, epithelium. Part of the basal membrane is fused in areas actively involved in gas exchange (*double-headed arrow*).

Direct lung lesions are most frequently due to lung trauma, such as contusion, aspiration, inhalation and burn injuries. Iatrogenic damage such as fluid overload and overtransfusion may also cause similar morphological alterations in the lungs.

Indirect lung injuries are mainly assigned to the broad pathological spectrum of shock (hypovolaemia, trauma, endotoxin, sepsis) and are related to the release of mediators with cytotoxic effects (see Chapter 3). The neurogenic interstitial lung oedema associated with craniocerebral trauma is also considered as indirect lung damage. It is probably caused by sympathico-adrenergic over-regulation.

Direct Lung Trauma

Contusion

One example of direct lung injury is pulmonary contusion as a consequence of trauma with or without rib fractures. Both the vascular system and the alveolar area are injured. The pathological changes are caused by damage to the alveoli and pulmonary capillaries, with subsequent interstitial and alveolar haemorrhage and oedema. The oedema and haemorrhage produce collapse of airways in the form of microatelectasis and segmental atelectasis. Surrounding these areas is a layer of less severely injured tissue with interstitial oedema and inflammatory cells (Mecca 1986).

Aspiration and Inhalation

Aspiration and inhalation damage primarily involves not only the proximal branches of the tracheobronchial tree but also the alveolar area and thus directly injures the gas exchange surface (Glause et al. 1979).

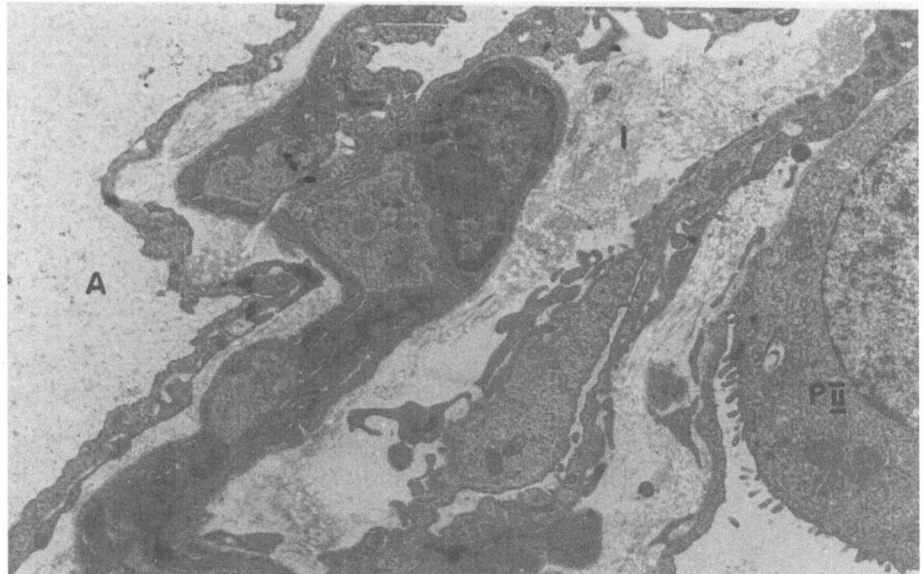

Fig. 2.2. Interstitial oedema (I) in a sheep lung 48 hours after inhalation injury. A, alveoli; PII, type II pneumocyte. ×5000.

Inhalation damage is often associated with progressive pulmonary oedema resulting in severe ARDS (Tranbaugh et al. 1983; Herndon et al. 1984). In experiments in sheep we were able to demonstrate—in addition to the many changes in haemodynamic and biochemical parameters (Traber et al. 1987)—a typical interstitial oedema (Fig. 2.2) in which damage to the endothelial capillary wall was not morphologically detectable. Currently it is not clear which specific mediators are responsible for the increase in permeability and from which morphological unit the resulting interstitial oedema originates. In this model the epithelial layer of the alveolar area is largely intact. Staining with esterase specific for polymorphonuclear leucocytes demonstrates an accumulation of granulocytes in the alveolar septa (Traber et al. 1984), as is also seen in the ultrastructure (Fig. 2.3).

Trauma-Related Indirect or Extrathoracic Lung Damage

It is well known that polytrauma leads to hypovolaemic-traumatic shock and as a consequence to decreased organ perfusion. This in turn, together with mediator release, may cause morphological and functional alterations in vital organs (e.g. the "lung in shock": Schlag et al. 1980). If appropriate treatment is instituted in time, the organ damage may be completely prevented or reversed. If treatment is delayed or inappropriate or if shock is severe, partially irreversible damage of the organ is the consequence and results in the shock syndrome in any one specific organ (e.g. shock lung syndrome).

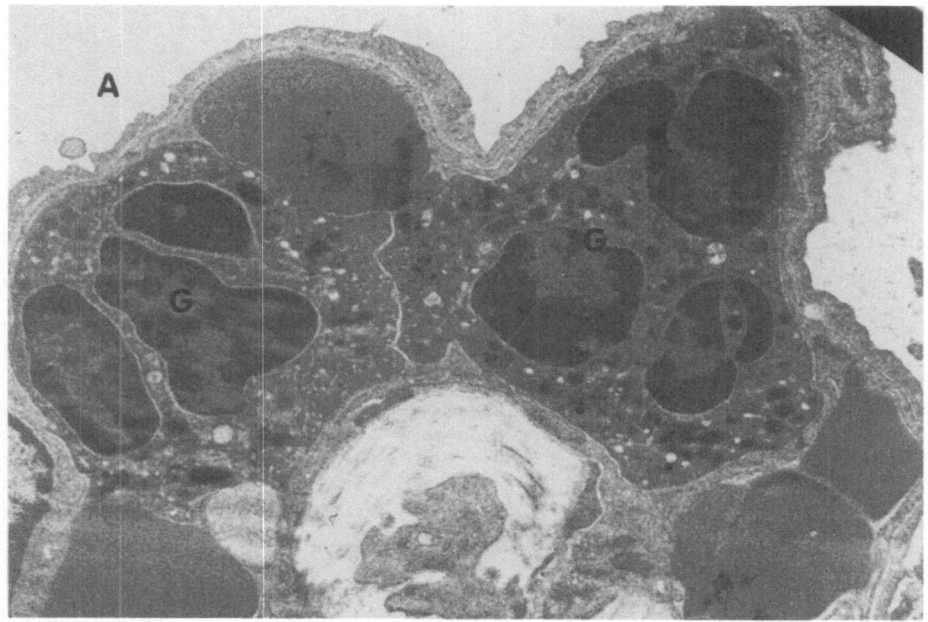

Fig. 2.3. Same experiment as in Fig. 2.2. Leucostasis (G, granulocytes) in pulmonary capillary. A, alveoli. ×5000.

The Lung in Shock

The most prominent ultrastructural alterations of the lungs during the initial stage of hypovolaemic-traumatic shock include the following:

Leucostasis with the accumulation of polymorphonuclear granulocytes (Fig. 2.3) that show partial degranulation with free lysosomal granules visible in the capillary lumen.

Variable swelling of the endothelial cells to the point of necrobiosis in some areas. Junctions are unaffected (Fig. 2.4).

Incipient interstitial oedema originating from around the microvasculature and spreading into the interstitial spaces (similar to that shown in Fig. 2.2).

Depending on the severity of coincident fractures and of the soft tissue injuries, isolated fat globules may be present.

In an attempt to obtain some experimental confirmation of these important ultra-structural changes which we observed in polytraumatized patients (Schlag and Redl 1985b), we developed models of hypovolaemic-traumatic shock in various animal species, but, in contrast to the human situation, under general anaesthesia. We demonstrated these changes of the "lung in shock" in the dog (Schlag and Redl 1985a), sheep and baboon (Pretorius et al. 1987). The morphological (ultrastruc-

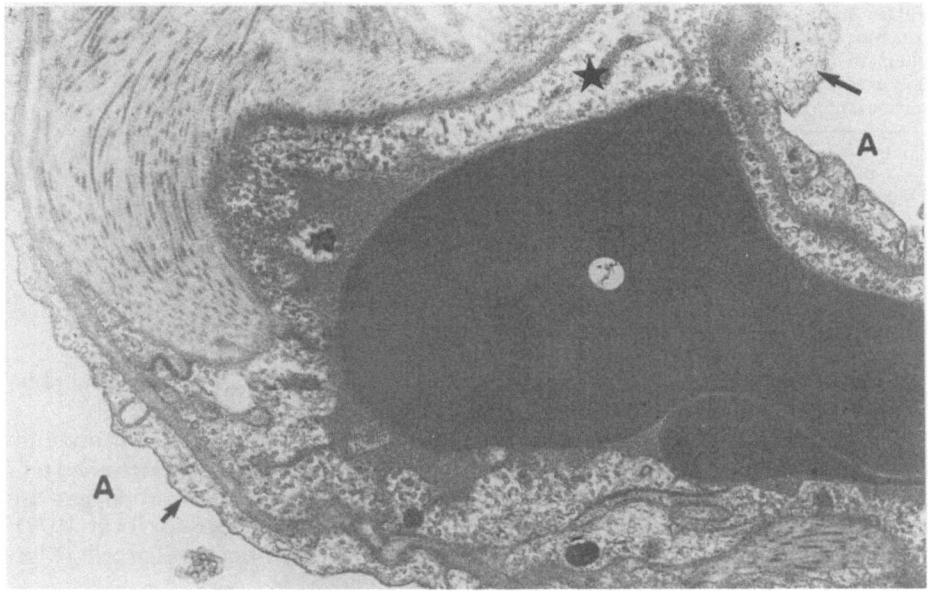

Fig. 2.4. Human lung biopsy after polytrauma with oedema of endothelial (*star*) and epithelial cells (*arrows*). A, alveoli. ×6500.

tural) changes of the lung in shock produced in our experimental models are very similar to those seen in the human lungs. Pulmonary leucostasis was dominant in all three species, while endothelial damage was less frequent in dogs and sheep but quite prominent in baboons. Interstitial oedema, however, was seen in all animals. Fat globules in the capillaries were rare.

Leucostasis in the lung is the morphological hallmark of shock and is always associated with severe polymorphonuclear leucocyte (PMN) degranulation. In accordance with these morphological findings, leucostasis was determined quantitatively (Redl et al. 1984a; Schlag and Redl 1985a) using ^{111}In-oxine-labelled neutrophils (Table 2.1). A recent study by Dinges et al. (1984) demonstrated a significant

Table 2.1. Leucostasis after polytrauma in dogs: quantitative evaluation using ^{111}In-oxine-labelled neutrophils (paired t-test)

	Control	Shock	
		Without reinfusion	With reinfusion
Pre-shock	2.50 ± 0.48	2.88 ± 0.46	1.76 ± 0.42
		$p < 0.05$	
Post-shock		7.02 ± 1.10	$p < 0.05$
Post-reinfusion	2.81 ± 0.19		5.58 ± 1.36
	$n = 6$	$n = 4$	$n = 4$

Table 2.2. Leucostasis after polytrauma in patients: quantitative evaluation using morphometric techniques after esterase staining of polymorphonuclear leucocytes in histological sections (unpaired t-test)

Control	Polytrauma
2.4 ± 0.9	8 ± 2.3
$n = 6$	$n = 6$
	$p < 0.001$

increase in PMNs in the lungs of patients who died within 48 hours following polytrauma, but did not in those with direct lung injury (Table 2.2).

The prerequisite of PMN recruitment in the lungs may be thought of as an accumulation after peripheral aggregation caused by C5a, low blood flow, an increased tendency to endothelial adherence or chemotaxis (e.g. alveolar macrophages are thought to be a source of neutrophil chemotactic activity: Kazmieroswski et al. 1977). There may also be the release of chemoattractant factors from endothelial cells (Mercandetti et al. 1984).

The granulocytes show a tendency to penetrate the interstitium from the capillaries of the lung. During PMN migration from the capillaries to the alveoli through the interstitium, degranulation occurs and this may cause additional interstitial damage secondary to release of proteinases and oxygen radicals.

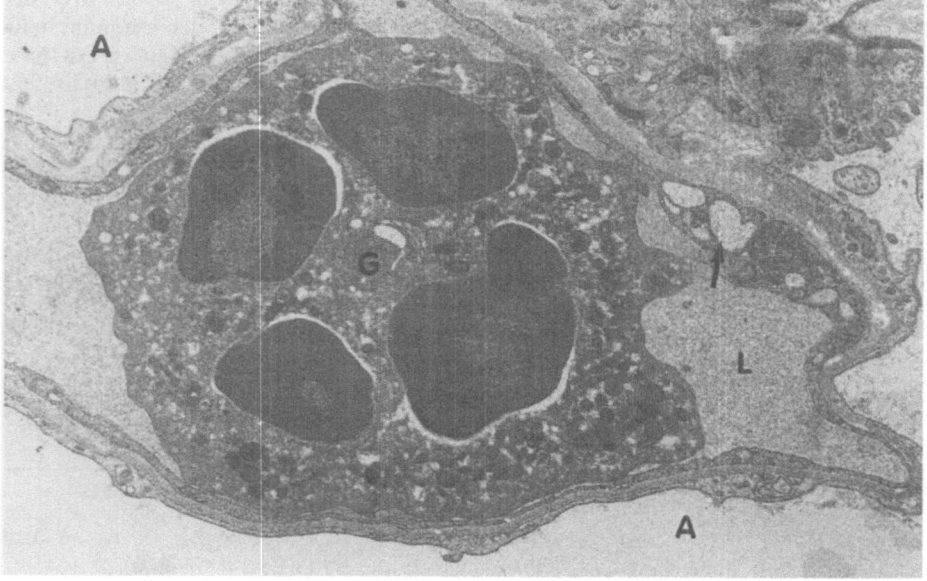

Fig. 2.5. Endothelial bleb formation (*arrow*) in post-traumatic lung of baboon. A, alveoli; G, granulocyte; L, capillary lumen. ×6500.

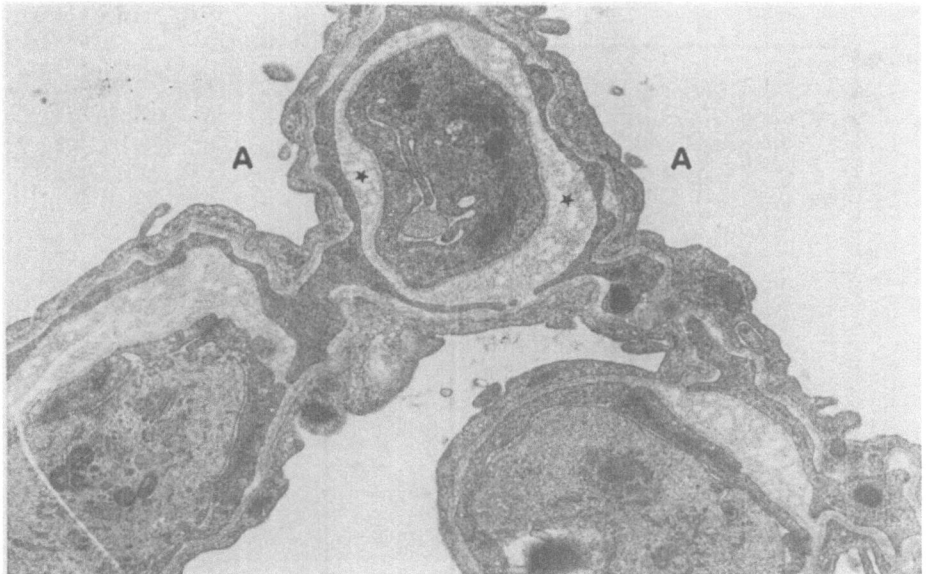

Fig. 2.6. Perivascular oedema (*stars*) of ovine lung after endotoxin infusion. A, alveoli. ×5000.

Other than the evidence from human biopsies (Schlag et al. 1976) the best demonstration of endothelial damage to the capillaries is in baboons, where endothelial bleb formation is found (Fig. 2.5). The bulla formation results from endothelial shedding. The endothelial protrusion can be caused by noxious stimuli—especially after vascular reperfusion following ischaemic periods (Gidlos et al. 1982; Hammersen and Hammersen 1984). Endothelial oedema is also the dominant finding in muscle biopsies from trauma patients (Schlag et al. 1977).

In all the species that we have studied, perivascular oedema (Fig. 2.6) is quite common and is found early on in the shock phase before reperfusion takes place. This is not clearly demonstrable by measurement of extravascular lung water (Redl et al. 1984b). However, after fluid replacement, the extravascular lung water increases significantly compared with the value in controls (Fig. 2.7).

However, fat globules in the capillaries of the alveolar septa are rare. They appear to be much more frequent in humans following an episode of traumatic shock.

Shock Lung Syndrome

If shock lung syndrome or post-traumatic ARDS is present, a distinction can be made between an acute and a subacute or chronic stage (Bachofen and Bachofen 1979). The shock lung syndrome shows the following structural alterations based on the morphological examination of the "lung in shock".

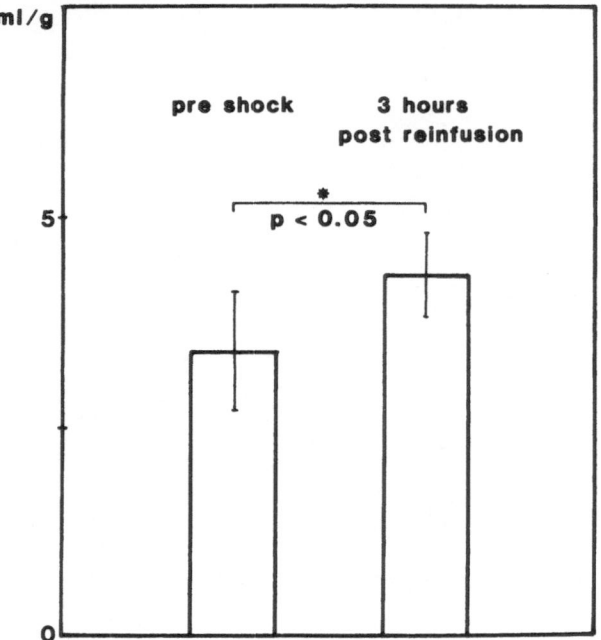

Fig. 2.7. Gravimetric extravascular lung water (open lung biopsies) per gram extravascular dry weight (EVLW/EVDW) in studies of canine hypovolaemic traumatic shock (duration of shock 1.5 hours). (P t-test, mean $\pm$ SD, $n = 6$.)

Acute Stage

In the acute stage interstitial and alveolar pulmonary oedema is widespread. In some patients the oedema fluid can contain fibrin, evident in studies with the electron microscope. This acute stage is often called the exudative phase.

The interstitial oedema is caused by increased permeability of the microvascular wall. In addition granulocytes pass through the swollen capillary endothelium to the interstitial space and alveolar areas. As the condition progresses and interstitial oedema becomes more prominent, alveolar epithelial cells tend to become oedematous (Fig. 2.8).

In areas of denuded alveolar cells (type I pneumocytes) increased permeability of the capillaries eventually leads to direct escape of interstitial oedema fluid into the alveolar space. Extravasation into the alveolar spaces of fibrin monomers—developed intravascularly—occurs simultaneously (Bleyl 1979). As these fibrin monomers become polymerized, hyaline membranes develop in the alveolar spaces (Fig. 2.9). These deposits may be related to an early inhibitory phase of fibrinolysis (Saldeen 1979). There is often a concomitant accumulation of leucocytes, erythrocytes and fat globules in the intra-alveolar space. These signal the presence of the "shock lung syndrome".

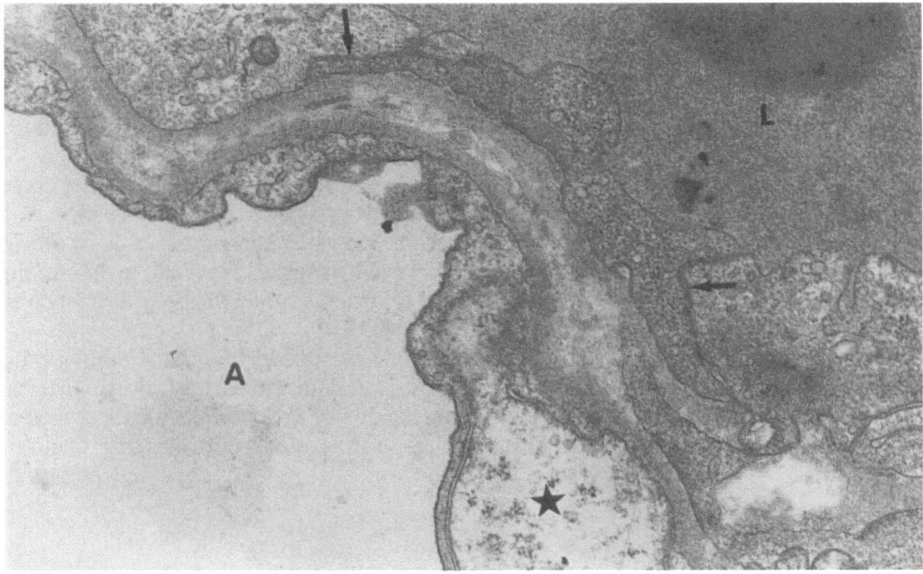

Fig. 2.8. Human lung capillary with focal oedema of endothelial cells (*arrows*) and epithelial cells (*star*). A, alveoli; L, lumen. ×10 000.

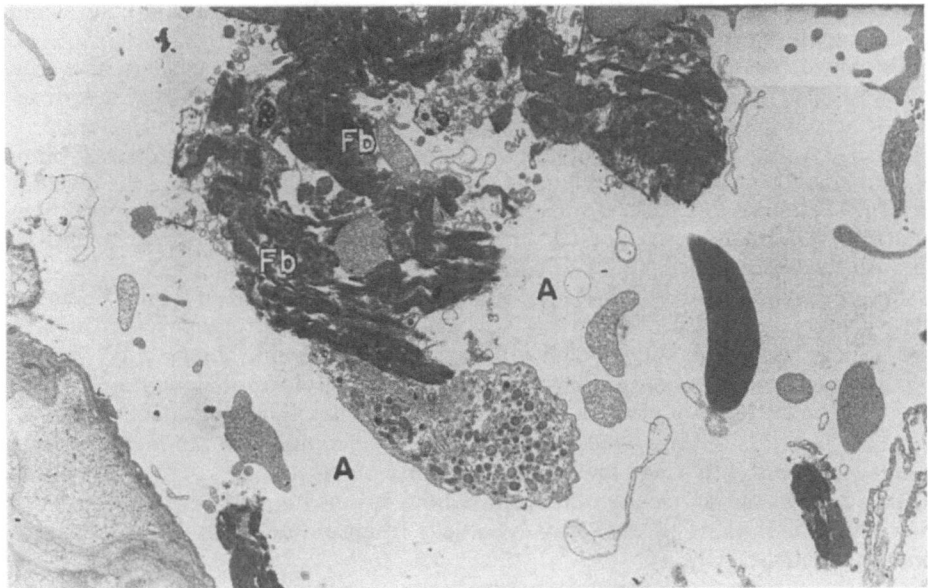

Fig. 2.9. Fibrin (Fb) during hyaline membrane formation in a human lung with ARDS. A, alveoli. ×3000.

To summarize the acute stage of ARDS, the morphological picture is of interstitial and alveolar pulmonary oedema with the oedema fluid containing high concentrations of fibrin. The alveolar epithelium is oedematous and has undergone partial destruction with denudation of the basement membrane. Type II pneumocytes which synthesize surfactant show empty vacuoles with most lamellar bodies expelled.

Macrophages in alveolar spaces frequently show the ultrastructural changes of pronounced vacuolization with large digestive vacuoles suggesting a macrophage overload. The clearing of the injured lung requires phagocytosis and digestion of a protein-rich, partly lipid-containing exudate as well as the degradation of fibrin deposits in the alveolar spaces. Thus the macrophage overload may lead to an impairment of pulmonary defence mechanisms. This may also reduce the host resistance to alveolar infection, a frequent complication of ARDS.

The acute stages of post-traumatic lung injury may be reversible if appropriate treatment is administered. However, in comparison with the earliest morphological phases, normalization of lung structure is a much slower process in the later stages of the disease. Progression to severe interstitial and intra-alveolar fibrosis may occur.

Subacute Stage

The subacute or chronic stage of ARDS is a proliferative process including fibrosis. Fibrosis may occur as early as the first week of any shock lung syndrome. Interstitial fibrosis is frequently associated with intra-alveolar fibrosis and impairment of gas exchange. The collagen content of the interstitium increases (Lamy et al. 1976). This is also visible in the ultrastructure (Schlag et al. 1985b) and has been chemically quantified by Zapol et al. (1979). The fibrosis is of prognostic significance and takes a rapidly progressive course. Shock-induced pulmonary fibrosis is reversible, but leaves isolated scars (Hassenstein et al. 1980). As a result, a complete structural and functional recovery is not possible.

Repair mechanisms such as a proliferation of type II pneumocytes are also seen (Thal et al. 1972; Teplitz 1976). Like fibrosis, these contribute to a further deterioration of gas exchange. Type II pneumocytes are also responsible for surfactant production. During the first few hours and days after injury their cytoplasm often appears disorganized, and lamellar bodies are discernible approaching the alveolar lumen. They seem to release these bodies very rapidly into the alveolar space (Schlag et al. 1980). In many cases type II pneumocytes then retain only clear vacuoles near their surface (Schlag and Redl 1985b).

Progressive interstitial fibrosis in conjunction with hyperplasia of type II pneumocytes leads to severe disturbances of gas exchange. Early and progressive fibrosis is of the utmost importance in the clinical outcome of acute lung injury and, when it occurs, is associated with a particularly poor prognosis. Fibrosis of later onset generally has a better prognosis and may be partially reversible (Hassenstein et al. 1980).

Acute progressive lung failure remains a dreaded complication of major trauma and is associated with a high mortality rate. Morphological studies of the early phase of the "lung in shock" clearly demonstrate how serious the immediate cellular and subcellular alterations are that may eventually result in the development of severe post-traumatic ARDS.

This paper is based on studies supported by grants of Lorenz Boehler Fonds, Nationalbankfonds, Fonds des Bürgermeisters der Bundeshauptstadt Wien.

References

Bachofen M, Bachofen H (1979) Der Heilungsverlauf des schweren "adult respiratory distress syndrome". Schweiz Med Wochenschr 109:1982–1989

Bleyl U (1979) Die Histophysiologie und Histopathologie der terminalen Lungenstrombahn bei akutem Lungenversagen. Klin Anaesthesiol Intensivmed 20:1–13

Dinges HP, Redl H, Schlag G (1984) Quantitative estimation of granulocyte in the lung after polytrauma: dog and human autopsy data. Eur Surg Res 16 [Suppl] 1:100–101

Gidlos A, Hammersen F, Larsson J, Lewis DH, Liljedahl SO (1982) Is capillary endothelium in human skeletal muscle an ischemic shock tissue? In: Lewis DH (ed) Induced skeletal muscle ischemia in man, Symposium Linköping, pp 63–79

Glause FL, Millen JE, Falls R (1979) Increased alveolar epithelial permeability with acid aspiration: the effects of high dose steriods. Am Rev Respir Dis 120:1119–1123

Hammersen F, Hammersen E (1984) The ultrastructure of microvascular endothelial cell reactions to various stimuli. Prog Appl Microcirc 6:91–108

Hassenstein J, Riede UN, Mittermayer C, Sandritter W (1980) Zur Frage der Reversibilität der schockinduzierten Lungenfibrose. Anaesthesiol Intensivther Notfallmed 15:340–349

Herndon DN, Traber DL, Niehaus GD, Linares HA, Traber LD (1984) The pathophysiology of smoke inhalation injury in a sheep model. J Trauma 24:1044–1051

Kazmierowski JA, Gallin JT, Reynolds HY (1977) Mechanism for the inflammatory response in primate lungs: demonstration and partial characterization of an alveolar macrophage derived chemotactic factor with preferential activity for polymorphonuclear leukocyctes. J Clin Invest 59:273–281

Lamy M, Fallat RJ, Koeniger E, Dietrich HP, Ratliff J, Eberhart RC, Tucker HJ, Hill JD (1976) Pathologic features and mechanisms of hypoxemia in adult respiratory distress syndrome. Am Rev Respir Dis 114:267–284

Mecca RS (1986) Pulmonary contusion and flail chest. Curr Rev Respir Ther 8:75–79

Mercandetti AJ, Lane TA, Colmerauer MEM (1984) Cultured human endothelial cells elaborate neutrophil chemoattractants. J Lab Clin Med 104:370–380

Pretorius JP, Schlag G, Redl H, Botha WS, Goosen DJ, Bosman H (1987) The "lung in shock" as a result of hypovolemic-traumatic shock in baboons. J Trauma (submitted for publication)

Redl H, Schlag G, Hammerschmidt DE (1984a) Quantitative assessment of leukostasis in experimental hypovolemic-traumatic shock. Acta Chir Scand 150:113–117

Redl H, Thurnher M, Krösl P, Schlag G (1984b) Extravaskuläre Lungenwasserbestimmung im hypovolämisch-traumatischen Schock (Frühpahse). Beitr Anaesthesiol Intensivmed 6:120–130

Saldeen T (1979) The microembolism syndrome: a review. In: Saldeen T (ed) The microembolism syndrome. Almqvist and Wiksell, Stockholm, pp 7–44

Schlag G, Redl H (1985a) Morphology of the microvascular system in shock: lung, liver, and skeletal muscle. Crit Care Med 13:1045–1049

Schlag G, Redl H (1985b) Morphology of the human lung after traumatic injury. In: Zapol WM, Falke KJ (eds) Acute respiratory failure. Marcel Dekker, New York Basel, pp 161–183

Schlag G, Voigt WH, Schnells G, Glatzl A (1976) Die Ultrastruktur der menschlichen Lunge im Schock: I. Anaesthesist 25:512–521

Schlag G, Voigt WH, Schnells G, Glatzl A (1977) Vergleichende Untersuchungen der Ultrastruktur von menschlicher Lunge und Skelettmuskulatur im Schock: II. Anaesthesist 26:612–622

Schlag G, Voigt WH, Redl H, Glatzl A (1980) Vergleichende Morphologie des posttraumatischen Lungenversagens. Anaesthesiol Intensivther Notfallmed 15:315–339

Teplitz C (1976) The core pathobiology and integrated medical science of adult acute respiratory insufficiency. Surg Clin N Am 56:1091–1131

Thal AP, Brown EB, Hermeck AS, Bell HH (1972) Shock: a physiologic basis for treatment. Year Book Medical Publishers, Chicago, pp 72–121

Traber D, Schlag G, Redl H, Traber L (1984) The mechanism of the pulmonary edema of smoke inhalation. Circ Shock 13:78

Traber DL, Schlag G, Redl H, Traber LD (1987) Pulmonary edema and compliance changes following smoke inhalation. Burn (submitted for publication)

Tranbaugh RF, Elings VB, Christensen JM, Lewis FR (1983) Effect of inhalation injury on lung water accumulation. J Trauma 23:597–604

Zapol W, Trelstad RL, Coffey JW, Tasi I, Salvador RA (1979) Pulmonary fibrosis in severe acute respiratory failure. Am Rev Resp Dis 119:547–554

Chapter 3

Mediators in Acute Lung Injury: The Whole Body Inflammatory Response Hypothesis

S. Westaby

A wide variety of extrapulmonary events such as major sepsis, burns, fat embolism, multiple trauma and pancreatitis may initiate an acute lung injury with the development of the adult respiratory distress syndrome (ARDS) (Anon. 1985). This same phenomenon, though usually in a much less severe form, has been observed in the controlled therapeutic situations of haemodialysis (Craddock et al. 1977), cardiopulmonary bypass (Westaby 1983) and nylon fibre leucapheresis (Nusbacher et al. 1978) where blood traverses a number of allegedly biocompatible foreign surfaces.

Although a large number of potential mediators have been identified in patients with ARDS, they differ according to the various aetiological problems and the interests of the researcher (Table 3.1). In the experimental animal model many of these agents can be demonstrated to cause the increased permeability of the alveolar capillary membrane that is characteristic of ARDS. In practical terms it is certain that the mechanisms of ARDS are far more complex than the action of a single agent on the pulmonary membranes. It is likely that diverse pathological processes initiate pulmonary membrane damage by the production of one of a small number of stimuli capable of triggering a common sequence of events (Fig. 3.1).

Recent observations suggest that the polymorphonuclear leucocyte (PMN) plays a central role in this process. Leucocytes are found sequestrated in the pulmonary capillaries of patients and experimental animals within an hour of the onset of haemorrhagic shock (Schlag et al. 1980). Increased numbers of PMNs are found in the broncho-alveolar lavage fluid from patients with ARDS (Cochrane et al. 1984) and Schlag et al. have described in detail the morphological changes in and ultrastructural damage to the alveolar capillary membrane as seen using electron microscopy (Schlag et al. 1980). The infusion of complement-activated plasma in rabbits causes an immediate systemic neutropenia sequestration of PMNs within the lungs followed by acute pulmonary oedema (Hohn et al. 1980). Infusion of phorbol-myristate acetate, a potent PMN stimulant, causes rapid pulmonary leucosequestration and gross morphological and histological evidence of an acute lung injury (Shasby et al. 1982). Depletion of circulating PMNs by nitrogen mustard prevents the effects induced by

Table 3.1. Potential mediators in acute lung injury

1. Complement components (C3a, C5a, C3b, C5b)
2. Kallikrein and kinins (bradykinin)
3. Products of coagulation (thrombin, fibrin aggregates, microthrombi)
4. Products of fibrinolysis (fibrin split products, fragment D, etc.)
5. Platelet derivatives (serotonin, thromboxane, etc.)
6. Vasoactive amines (histamine, serotonin)
7. Products derived from arachidonate:

    (a) Non-enzymatically   – Oxidation products of arachidonate
    (b) Via cyclo-oxygenase   – Endoperoxides (PGG_2, PGH_2)
                                     Thromboxane
                                     Prostacyclin (PGI_2)
    (c) Via lipoxygenase   – Hydroxyperoxy acids (HPETE)
                                     Hydroxy acids (HETE)
                                     Leukotrienes

8. Neutrophil derivatives:

    (a) Lysosomal constituents   – Elastase
                                     Collagenase
                                     Myeloperoxidase
                                     Cathepsins
                                     Lysozyme
    (b) Oxygen free radicals   – O_2^-, H_2O_2, $OH^{\cdot}$
    (c) Arachidonic acid metabolites – As in 7 above

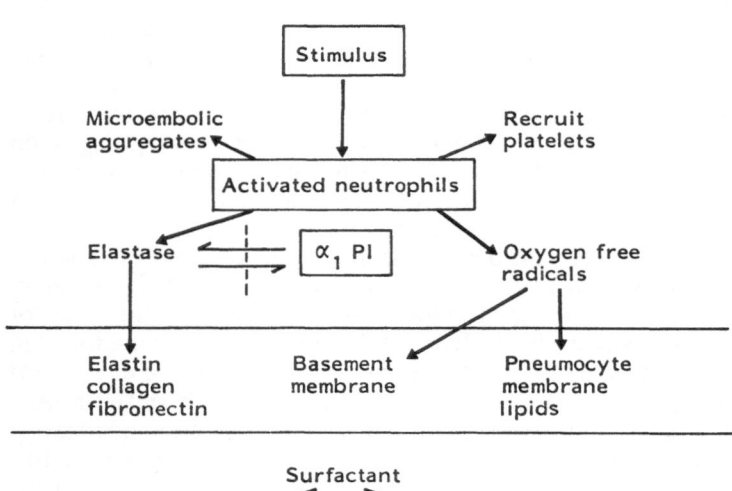

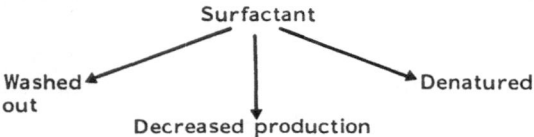

Fig. 3.1. Hypothetical sequence of events for neutrophil activation, pulmonary membrane damage and surfactant depletion in acute lung injury. α_1PI, α_1 protease inhibitor.

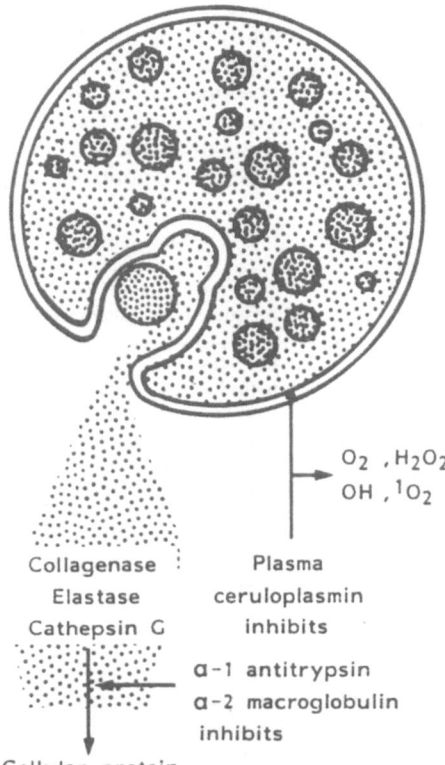

O_2, H_2O_2
OH, $^1O_2$

Collagenase
Elastase
Cathepsin G

Plasma
ceruloplasmin
inhibits

$a-1$ antitrypsin
$a-2$ macroglobulin
inhibits

Cellular protein

Fig. 3.2. Release of lysosomal enzymes and generation of oxygen free radicals by "activated" polymorphonuclear leucocytes.

phorbol-myristate acetate to the extent that there is no discernible lung damage. In clinical studies in man, pulmonary leucosequestration has been demonstrated in the transient pulmonary dysfunction associated with haemodialysis (Craddock et al. 1977) and cardiopulmonary bypass (Chenoweth et al. 1981), and this is consistent with the finding that PMNs possess several different mediator systems capable of altering membrane structure and function (Westaby 1986) (Fig. 3.2).

The Whole Body Inflammatory Response Hypothesis

Most clinical studies in ARDS are of necessity retrospective, monitoring the end stage of the disease process and not the pathogenesis. The parallel between ARDS following trauma or sepsis and the pulmonary dysfunction of extracorporeal circulation has allowed certain investigations to be performed in a prospective manner. Cardiopulmonary bypass is followed by diffuse systemic effects which in an exaggerated and severe form have been known as the "post-perfusion syndrome". The lungs most frequently manifest signs of damage which morphologically and functionally resemble the changes in ARDS (Asada and Yamaguchi 1971; Ratliffe et al. 1973). In addition, renal dysfunction, fever of non-infective origin, diffuse interstitial oedema and leucocytosis suggest a generalized condition which in certain respects resembles the acute inflammatory response at the site of a local injury. This has led to the hypothesis that such changes represent a "whole body" inflammatory reaction against the foreign materials of the extracorporeal circuit, with tissue damage occurring by the wide dissemination of activated neutrophils throughout the body (Table 3.2).

Table 3.2.   Concept of the whole body inflammatory response

Local injury or sepsis	Mediators and inflammation restricted to the local site, humoral cascades controlled by circulating inhibitors **Chemotaxis brings white cells to the site**
Polytrauma or major sepsis	Inhibitors overwhelmed. Widespread activation of humoral cascades. Mediators released into the circulation. **Systematic dissemination of activated white cells**

Complement activation had previously been implicated in the pulmonary dysfunction associated with haemodialysis (Craddock et al. 1977) and this has been investigated during cardiopulmonary bypass using the highly sensitive C3a and C5a anaphylatoxin assays (Kirklin et al. 1983).C3a was released into the circulation of all patients undergoing surgery with cardiopulmonary bypass, but not in patients undergoing other cardiovascular procedures under general anaesthesia who did not receive cardiopulmonary bypass. Plasma C3a levels 3 hours after the end of perfusion were related to the probability of abnormal bleeding and to cardiac, pulmonary and renal dysfunction in the postoperative period. At the end of cardiopulmonary bypass the reversal of heparinization by the infusion of protamine causes further complement activation. This has been associated with pronounced cardiovascular changes and anaphylactoid reactions with haemorrhagic pulmonary oedema (Westaby et al. 1984a).

The half-life of C5a anaphylatoxin in the circulation is extremely short due to its rapid binding of PMNs, monocytes and other plasma membranes. Measurement in plasma during cardiopulmonary bypass fails to demonstrate elevated levels, but at reperfusion of the pulmonary vascular bed at the end of bypass, PMNs and monocytes (but not lymphocytes or eosinophils) are temporarily sequestrated within the microcirculation (Gavasocchi et al. 1986). This trapping coincides with evidence of free radical peroxidation of membrane lipids and the release of PMN proteolytic enzymes such as elastase (Westaby et al. 1984b). The result is altered alveolar capillary membrane function followed by non-cardiogenic pulmonary oedema (Royston et al. 1985). This sequence clearly parallels local inflammation by some stimulus leading to PMN and monocyte activation and accumulation, and the ability to study perfusion-related damage prospectively has helped to shed light on the likely mechanism of damage after other traumatic and septic events.

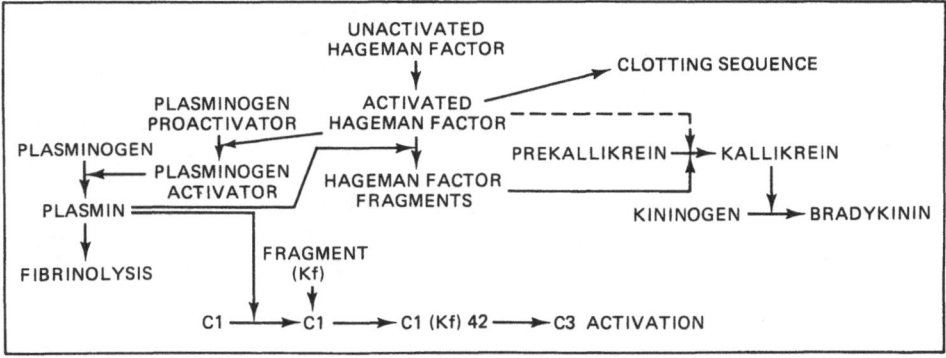

Fig. 3.3. Interrelationship between the humoral cascades: complement, coagulation, fibrinolysis and kallikrein.

Mechanisms of White Cell Activation in ARDS

The inflammatory response to local injury or sepsis results from the homeostatic mechanisms of one or more closely integrated humoral cascades—complement, coagulation, fibrinolysis and kallikrein (Ruddy et al. 1972) (Fig. 3.3). Activation of one or more cascade systems produces byproducts many of which have been implicated as mediators in ARDS (Fig. 3.4). For instance, complement produces the vasoactive chemotaxins C3a and C5a while coagulation and fibrinolysis produce thrombin which damages endothelial cells directly, fibrinopeptides which cause pulmonary vasoconstriction, and fibrin monomers which stimulate arachidonic acid metabolism, thromboxane release and pulmonary vasoconstriction. Fragment D produces neutrophil chemotaxis and has been implicated in the pulmonary dysfunction of major burns (Luterman et al. 1977).

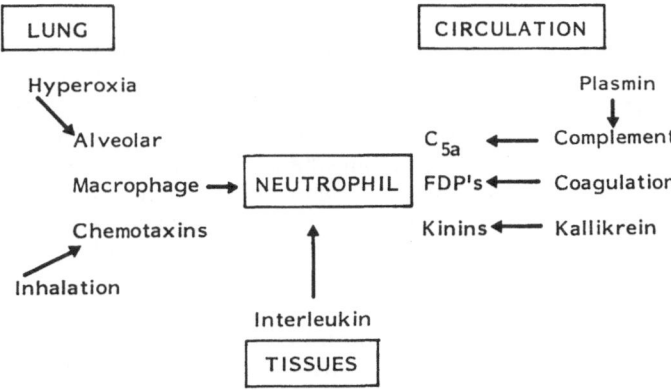

Fig. 3.4. Stimulus for neutrophil "activation" in trauma, sepsis and extracorporeal circuits.

The humoral cascades are regulated by the circulating antiprotease inhibitors (α_1 protease inhibitor, α_2 macroglobulin, C1 inactivator and antithrombin III) which prevent widespread catastrophic propagation. The result is the contained liberation of those factors and others (such as the various different monokines) from traumatized tissue which attract PMNs to the site and initiate the acute inflammatory response. Bacteria and debris are then dealt with by phagocytosis and intracellular digestion by proteolytic enzymes.

In the face of massive tissue injury, major burns and severe sepsis, widespread activation of the humoral cascades may overwhelm the circulating inhibitors resulting in systemic effects including PMN activation. This is the setting for a whole body inflammatory reaction in which white cells (activated PMNs and monocytes) are trapped in the vascular bed of the lungs, kidneys and other organs.

The predilection for intrapulmonary trapping of white cells is not completely understood. Certainly the lung contains an extensive microcapillary bed which may act as a filter for aggregates in venous blood. C5a causes PMN aggregation and an aggregate of 30–40 cells produces an embolus 80–90 µm in diameter. However, it is possible that PMNs are actively recruited to the lungs (Reynolds 1983). The alveolar macrophage is able to elaborate chemotaxins and it has been shown that hyperoxia-stimulated alveolar macrophages produce distinct factors that encourage neutrophil adhesion and neutrophil superoxide generation (Harada et al. 1982). During cardiopulmonary bypass it is likely that the bypassed (ischaemic) lung produces tissue plasminogen activator which in turn activates complement with the local generation of C5a. Experimental endotracheal instillation of C5a causes the accumulation of PMNs, alterations in alveolar epithelium and protein-rich alveolar oedema (Larsen et al. 1980). In PMN-depleted animals C5a instillation does not cause the same structural alterations of permeability changes.

Mechanisms of Pulmonary Membrane Damage by Neutrophils

Pulmonary leucostasis has great importance in subsequent events leading to membrane damage. Neutrophils contain many lysosomes with powerful hydrolytic and proteolytic potential. Normally the lysosomal enzymes along with oxidizing agents (free radicals) produced by phagocytosis serve two main purposes (Johnson et al. 1975): firstly, intracellular protein catabolism including the degradation of intracellular and extracellular endogenous substances; and secondly, the degradation of phagocytosed viruses and bacteria. These functions are normally fulfilled within the cell by the formation of a phagosome. Henson has suggested a mechanism for neutrophil-mediated tissue injury by proposing that neutrophils might be "frustrated" in their attempts to ingest impossibly large materials such as damaged endo- and epithelial surfaces with adherent immune complexes (Henson 1972). The inability of neutrophils to phagocytose such particles might result in the spillage of toxic neutrophil products into the extracellular space with subsequent host tissue injury. Neutrophils may release at least three groups of products that could cause tissue damage. These are lysosomal proteolytic enzymes, toxic oxygen free radicals and products of arachidonate metabolism (Tate and Repine 1983).

Neutrophil granules contain a variety of different proteins including elastase, collagenase, cathepsins, lysozyme, lactoferrin and myeloperoxidase (Gnanaduoai et al. 1978). Many of these substances can damage normal cells and tissues. Elastase and collagenase can digest basement membrane, elastic tissue in arterial walls and pulmonary elastin. In addition to degrading lung tissue components, PMN proteases can amplify the inflammatory process by cleaving complement, fibrinogen and Hageman factor.

Similarly toxic oxygen species have been shown to damage a number of different cell types including fibroblasts, lung parenchymal cells, endothelial cells and erythrocytes (Martin et al. 1981). There are several mechanisms for this damage. Key cellular components such as enzymic proteins and membrane lipids are highly susceptible to oxidation. Protective enzymes such as superoxide dismutase or catalase may not be in the right location to protect from oxidant injury. Oxidation of unsaturated fatty acids (such as linoleic acid and arachidonic acid) forms lipid peroxide radicals and initiates a chain reaction between lipid peroxide and membrane fatty acids. This chain reaction can be interrupted by antioxidants such as vitamin E (Harada et al. 1983). Lipid peroxide radicals also decompose to form malondialdehydes which can cross-link and damage free amino groups of proteins, nucleic acids and phospholipids. Increased levels of malondialdehyde have been measured during intrapulmonary white cell sequestration during cardiopulmonary bypass (Royston et al. 1986).

Neutrophils can produce and release a spectrum of arachidonate-derived products such as prostaglandins, thromboxanes and leukotrienes. These substances have major effects on pulmonary vascular smooth muscle and capillary permeability (Brigham and Ogletree 1981). Infusion of arachidonic acid or related products raises pulmonary vascular resistance. Products of the lipoxygenase pathway, including leukotrienes A4, B4 and the slow-reacting substance of anaphylaxis—leukotrienes C4, D4 and E4—have been shown to participate in inflammatory and oedemogenic processes.

The inter-relationships between the PMN-derived agents are complex (Schraufstatter et al. 1984). The principal inhibitor of PMN elastase is α_1 protease inhibitor, the activity of which is seriously compromised by toxic oxygen species (Zaslow et al. 1983). Sequestrated white cells may therefore also destroy the defence mechanisms against proteolytic enzyme release.

Conclusion

A great many different mediators have been purported to cause pulmonary membrane damage in ARDS. The "whole body inflammatory response" hypothesis offers an explanation for the diverse cellular and humoral events known to occur in this condition. Investigations during clinical cardiopulmonary bypass and in the animal laboratory lend support to this hypothesis and form a basis on which logical intervention may be planned.

References

Anonymous (1985) Adult respiratory distress syndrome. Lancet I:301–303

Asada S, Yamaguchi M (1971) Fine structural changes in the lung following cardiopulmonary bypass. Chest 59:478

Brigham KL, Ogletree ML (1981) Effects of prostaglandins and related compounds on lung vascular permeability. Bull Eur Physiopathol Respir 17:703–722

Chenoweth DE, Cooper SW, Hugli TE, Stewart R, Blackstone EH, Kirklin JW (1981) Complement activation during cardiopulmonary bypass: evidence for generation of C_{3a} and C_{5a} anaphylatoxins. N Engl J Med 304:497

Craddock PR, Fehr J, Dalmasso AP, Brigham KL, Jacobs HS (1977) Haemodialysis leukopenia. Pulmonary vascular leukostasis resulting from complement activation by dialysis cellophane membrane. J Clin Invest 59:589

Cochrane CG, Spragg RG, Davat SD, Cohen AB, McGuire WW (1984) The presence of neutrophil elastase and evidence of oxidant activity in broncho-alveolar lavage fluid of patients with adult respiratory distress syndrome. Am Rev Respir Dis [Suppl] 127:25–27

Gavasocchi NC, Pluth JR, Schaff HV, Orszulak TA, Homburger HA, Solis E, Kaye MP, Clancy MS, Kolff J, Deeb GM (1986) Complement activation during cardiopulmonary bypass: comparison of bubble and membrane oxygenators. J Thorac Cardiovasc Surg 91:252–258

Gnanaduoai TV, Branthwaite MA, Colbeck JF, Wellman E (1978) Lysosomal enzyme release from the lungs after cardiopulmonary bypass. Anaesthesia 33:227

Harada RN, Bowman CM, Fox FB, Repire JE (1982) Alveolar macrophage secretions: initiators of inflammation in pulmonary oxygen toxicity. Chest 81:52–54

Harada RV, Vatter AE, Repine JE (1983) Oxygen radical scavengers protect alveolar macrophages from hyperoxic injury in vitro. Am Rev Respir Dis 128:552–559

Henson RM (1972) Pathologic mechanisms in neutrophil-medicated injury. Am J Pathol 68:593–612

Hohn DC, Meyers AJ, Gherini ST, Beckman A, Markison RE, Churg AM (1980) Production of acute pulmonary injury by leukocytes and activated complement. Surgery 88:48

Johnson RB, Keele BB, Misoa HP (1975) The role of superoxide anion generation in phagocytic bactericidal activity: studies with normal and chronic granulomatous disease leukocytes: J Clin Invest 55:1357–1372

Kirklin JK, Westaby S, Blackstone EH, Kirklin JW, Chenoweth DE, Pacifico AD (1983) Complement and the damaging effects of cardiopulmonary bypass. J Thorac Cardiovasc Surg 86:845

Larsen GL, McCarthy K, Webster RO, Henson JE, Henson PM (1980) A differential effect of C_{5a} and C_{5a} des arg in the induction of pulmonary inflammation. Am J Pathol 100:179–192

Luterman A, Manwaring D, Curreri PW (1977) The role of fibrinogen degradation products in the pathogenesis of the respiratory distress syndrome. Surgery 82:703–709

Martin WJ, Gadek JE, Hunninghake GW, Crystal RG (1981) Oxidant injury of lung parenchymal cells. J Clin Invest 68:1277–1288

Nusbacher J, Rosenfeld SI, MacPherson JL, Theim PA, Leddy JP (1978) Nylon fibre leukophoresis: associated complement component changes and granulocytopenia. Blood 51:359

Ratliffe NB, Young WA, Hackel DB, Mikat E, Wilson JW (1973) Pulmonary injury secondary to extracorporeal circulation. J Thorac Cardiovasc Surg 65:425

Reynolds HY (1983) Lung inflammation: role of endogenous chemotactic factors in attracting polymorphonuclear granulocytes. Am Rev Respir Dis 127 [Suppl]:S16–S25.

Royston D, Minty BD, Higgenbottam TW, Wallwork J, Jones GJ (1985) The effect of surgery with cardiopulmonary bypass on alveolar-capillary barrier function in human beings. Ann Thorac Surg 40:139–143

Royston D, Fleming JS, Westaby S, Taylor KM (1986) Increased production of peroxidation products associated with open heart surgery: evidence for free radical generation. J Thorac Cardiovasc Surg (in press)

Ruddy S, Gigli I, Austen KF (1972) The complement system of man. N Engl J Med 287:489–495

Schlag G, Voigt WH, Redl H, Glatzl A (1980) Vergleichende Morphologie des posttraumatisches Lungenversagens. Anaesthesiol Intensivther Notfallmed 15:315–339

Schraufstatter IU, Revak SD, Cochrane CG (1984) Proteases and oxidants in experimental pulmonary inflammatory injury. J Clin Invest 73:1175

Shasby DM, Van Benthuysen KM, Tate RM, Shasby SS, McMustry IF, Repire JE (1982) Granulocytes mediate acute oedematous lung injury in rabbits and isolated rabbit lungs perfused with phorbolmyristate acetate: role of oxygen radicals. Am Rev Respir Dis 125:443–447

Tate RM, Repine JE (1983) Neutrophils and the adult respiratory distress syndrome. Am Rev Respir Dis 128:552–559

Westaby S (1983) Complement and the damaging effects of cardiopulmonary bypass. Thorax 38:321

Westaby S (1986) Mechanisms of membrane damage and surfactant depletion in acute lung injury. Intensive Care Med 12:2–5

Westaby S, Turner MW, Stark J (1984a) Complement activation, haemorrhagic pulmonary oedema and peripheral circulatory collapse following protamine administration in a child after cardiopulmonary bypass. Br Heart J 53:574–576

Westaby S, Fleming J, Royston D (1984b) Evidence for generation of free radical species during cardiopulmonary bypass in man. In: Hagl S, Klovekorn WP, Mayr N, Sebering F (eds) Thirty years of extracorporeal circulation. Deutches Herzzentrum, Munich, p 389

Zaslow MC, Clark RA, Stone PJ, Calose JD, Snider GL, Franzblau C (1983) Human neutrophil elastase does not bind to alpha$_1$-protease inhibitor that has been exposed to activated human neutrophils. Am Rev Respir Dis 128:434–439

Chapter 4

Intravascular Microaggregates and Pulmonary Embolization in Shock and Surgery

C. N. McCollum and K. R. Poskitt

Adult respiratory distress syndrome (ARDS) is a frequent cause of death in intensive care units and probably represents a collection of different conditions. In the complex issue of its aetiology the failure to define terms has led in our opinion to considerable confusion. It is doubtful whether the pulmonary failure that follows viral pneumonia or specific lung infections has the same pathophysiology as that following 3 days after massive blood transfusion, although both may share similar X-ray features and be called "ARDS". In this chapter "shock" refers to that condition in which there is a failure of the circulation of blood to meet the metabolic needs of the tissues. "Shock lung" is defined here as pulmonary failure without left ventricular failure typified by increased pulmonary arteriovenous admixture and widespread opacification of the lung fields on X-ray, developing as a consequence of and at an interval of hours or days following an episode of shock or severe sepsis. In this way we may concentrate entirely on that variety of ARDS that almost certainly has a common, although possibly multiple, cause that was almost certainly active during the episode of shock. The delay before pulmonary failure develops is a consistent feature of great importance in understanding the pathophysiology.

Shock and Shock Lung

In the early 1960s the main problem in the management of shock was perceived to be the development of "irreversible shock" which was attributed to excessive sympathetic activity, stagnation in the microcirculation and diffuse intravascular coagulation or platelet aggregation (Hardaway et al. 1961; Robb 1963; Lillehei et al. 1964).

It was then appreciated that vasoconstricting drugs were damaging and that blood flow is more important than blood pressure for tissue perfusion. The term "irreversible shock" was then virtually abandoned and increasing numbers of patients survived severe shock only to develop the lethal complications of pulmonary, renal or hepatic failure (Moore et al. 1969). It seems likely that similar processes are responsible for failure of all these vital organs during shock, sepsis and following major surgery.

A consistent feature of shock lung is that the patient may be well for 24–48 hours following the episode of shock, with no signs of pulmonary dysfunction other than tachypnoea (Moore et al. 1969). Pulmonary failure becomes clinically apparent on the second to sixth day, with an early presentation indicative of a poor prognosis. Increasing dyspnoea, central cyanosis and diffuse opacification on the X-ray herald massive pulmonary venous admixture and pulmonary failure. As hypoxia is based on pulmonary shunting with increasing alveolar–arterial PO_2 gradients, positive pressure ventilation with high inspired oxygen tensions produces little benefit (Del Guercia et al. 1966; Blaisdell and Lewis 1977). Once the full features of shock lung have developed the outlook is dismal, with a mortality of over 90% reported in a multi-centre study attempting the use of extracorporeal membrane oxygenation (Murray 1977).

As the prognosis in established shock lung is poor and there are no new therapeutic approaches that are likely to influence this, our attention must be directed towards prevention. Prophylaxis, however, demands an understanding of the aetiology, an issue that is not yet resolved.

The Aetiology of Shock Lung

The range of suggested aetiologies for this condition is so great that there would be little purpose in even listing them all. Those that gained some degree of acceptance during the last two decades are shown in Table 4.1 to illustrate the extent of controversy. The main difficulty has been that all these conditions can potentially cause pulmonary failure with features similar to those we see in shock lung. The real question is which pathology or combination of pathologies both develops in shock and causes the subsequent pulmonary damage.

Table 4.1.   "Causes" of shock lung

Pulmonary ischaemia	Bergofsky (1970)
Neurogenic factors	Simmons et al. (1968)
Oxygen toxicity and ventilation	Nash et al. (1971)
Fluid overload	Collins (1969)
Toxins	Lillehei et al. (1964), Collins (1978)
Humoral factors	Clowes et al. (1970), Cook and Webb (1968)
Pulmonary embolization	
Fat embolism	Fuchsig et al. (1967)
Transfused aggregates	Hissen and Swank (1965), McNamara et al. (1972)
Leucocyte sequestration	Craddock et al. (1977), Jacob (1978)
Endogenous aggregates	Blaisdell et al. (1970), McCollum and Campbell (1979)

The Evidence for Pulmonary Microemboli

Following the many studies that have been performed on this condition it is now recognized that excluding the lung, or part of the lung, from the circulation during shock protects it from damage (Nahas et al. 1965; Shah-Mairany et al. 1970). Furthermore, pulmonary damage is most prominent in the well-perfused areas of the lung (Tiefunbrun et al. 1975). The damaging agent is therefore carried to the lungs by the circulation and it is attractive to suppose that it embolizes causing damage to the microcirculation. Although fat embolism can be demonstrated in a small proportion of patients it is now accepted that it is absent in the majority and is unlikely to be the main aetiological factor in patients without the recognizable and rare fat embolism syndrome that occurs following trauma. One of the few universally accepted observations relating to all varieties of shock is that there is an immediate fall in the platelet and leucocyte counts. This led to the concept that these cells may aggregate and embolize to the lungs as a result of some stimulus during the episode of shock. Platelets and possibly leucocytes appear essential in promoting pulmonary damage in that their removal prior to an episode of shock protects pulmonary function (Stein and Thomas 1967; Bredenberg et al. 1980).

Even if it is accepted that extensive pulmonary embolization from platelets, leucocytes or a combination of both may be the cause of pulmonary failure, it is not clear whether this is initiated by activated complement as suggested by Craddock et al. (1977), Jacob (1978) and Hammerschmidt et al. (1980). Also, the influence of exogenous aggregates derived from blood transfusion is uncertain, but these aggregates do have the potential of adding to pulmonary dysfunction by promoting pulmonary arteriovenous shunting (Geelhoed and Bennett 1975; Brown et al. 1977). Even in the absence of blood transfusion fibrin and platelet aggregates may be demonstrated in the pulmonary vasculature following haemorrhagic or endotoxic shock (Lim et al. 1966; Harrison et al. 1977). Death of the patient may not occur until some time after the aggregates have lysed or been phagocytosed, but microscopy of the lungs within 48 hours of fatal injury has demonstrated platelet aggregates in the microcirculation (Blaisdell et al. 1970; Moseley and Doty 1970). In our own work we have used major surgery as a model for shock in patients and demonstrated a clear relationship between aggregates measured in the venous circulation at the time of surgery and subsequent pulmonary dysfunction (McCollum and Campbell 1979).

Major Surgery as a Shock Model in Man?

In the immediate postoperative period following major surgery most patients appear clinically to be shocked. Even where the operative fluid requirements have been estimated accurately and the blood pressure is normal there is usually tachycardia, cold extremities with peripheral cyanosis and often a poor urinary output persisting for some hours. In many procedures endotoxin may be detectable in portal and peripheral blood (Bailey 1976). As in shock, pulmonary failure is the most frequent cause of death following major surgery. This pulmonary failure is typified by venous admixture and as such is indistinguishable from "shock lung" and may well have an identical aetiology.

Coagulation Disorders in Shock and Surgery

Hewson (1772) found that when haemorrhage occurs in an animal "the blood to issue first clots last". This hypercoagulability of shock has been recognized for over two decades and was believed to be one of the important causes of "irreversible shock" (Hardaway 1962; Attar et al. 1966). In both shock and following surgery the first detectable change is a rapid and often precipitous fall in the platelet count (Hardaway 1962; Ygge 1970). This fall in platelet count is followed by a fall in fibrinogen levels and a rise in fibrinogen degradation products. These changes in coagulation parameters are identical for shock and surgery and in both are followed by activation of the fibrinolytic system consuming plasminogen and other fibrinolytic factors. In the postoperative and post-shock period the capacity for lysis is then reduced and there is an increase in the incidence of thromboembolic events.

These similarities between the consequences of major surgery and shock on blood coagulation are hardly surprising as many shock patients undergo surgery and many postoperative patients develop complications causing shock. If intravascular cellular aggregation is a cause of shock lung then major surgery may be the optimal model for its study in man. Our first requirement is a precise method of measuring circulating intravascular aggregates.

The Measurement of Intravascular Aggregates

Of the several possible techniques for detecting cellular aggregates in shocked patients a modification of the screen filtration pressure technique seems most appropriate (Swank 1961). Following pilot studies we used a nickel screen filter of pore size 20 μm square and deep which was sealed in a filter block so that the diameter exposed to blood flow was 2.4 mm (Fig. 4.1). The technique has been described

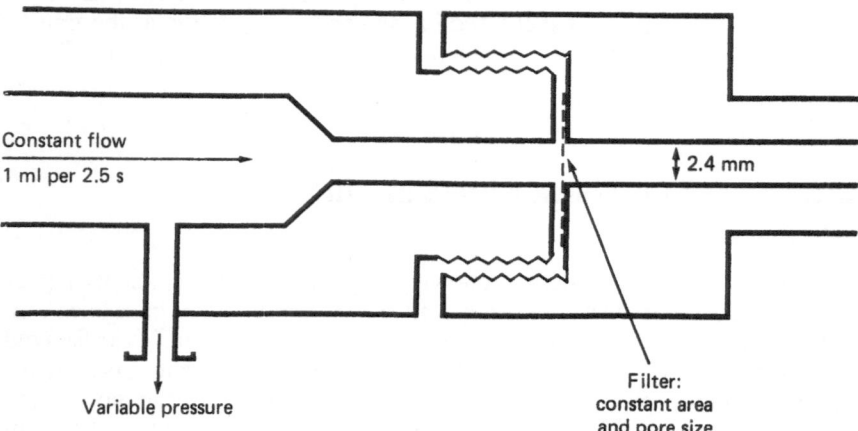

Fig. 4.1. In screen filtration pressure measurement, 5 ml of femoral vein blood is passed at constant flow through a constant area of nickel screen filter. Pressure proximally will rise, with the rate of rise reflecting the number of aggregates occluding the filter pores.

in detail whereby the pressure proximal to the filter is measured by transducer and reflects the resistance to flow through the filter (McCollum and Campbell 1979). In essence, a saline sample is passed first to calibrate the apparatus and to remove all air, followed by the test sample of a constant volume of the patient's femoral vein blood. A further sample of saline is then passed immediately to wash excess blood from the filter surface prior to its examination by scanning electron microscopy.

Using this technique we found in 80 consecutive estimations on postoperative patients that the slope of the pressure rise during the passage of the test sample of blood most closely correlated with the number of aggregates seen on the filter surface by scanning electron microscopy (correlation coefficient $r = 0.93$, $p<0.001$). The reproducibility of the test was excellent as, unlike the peak pressure obtained, the slope of the screen filtration pressure (SFP) curve is not influenced by viscosity. We found significantly more aggregates in blood from the femoral vein than from the basilic vein, suggesting that the combination of poor peripheral perfusion with venous stasis may be important in their formation.

Prior to detailed studies on the effects of these aggregates on the lungs we examined their prevalence in a group of 43 patients undergoing either minor surgery such as hernia repair or major surgery such as aortic replacement or pancreatic or colonic resection. Femoral vein blood was evaluated as described above within 20 minutes of the completion of surgery. Preoperative measurements were performed in 36 of the 43 patients and in 35 of these the SFP was below 5 (Fig. 4.2). Similar results were obtained in 14 patients following minor surgery with a mean SFP ($\pm$SD) slope of 1.5 $\pm$ 1.6. Of the 29 estimations in patients following major surgery, SFP was greater than 5 in 17 with a mean for this group of 7.1 $\pm$ 6.1. This was markedly greater than that either preoperatively or following minor surgery ($p<0.001$). The normal range for the technique was calculated by adding twice the standard deviation to the mean for measurements preoperatively and following minor surgery, giving an upper limit of normal calculated at 4.7. SFP measurements of 5 or greater were therefore

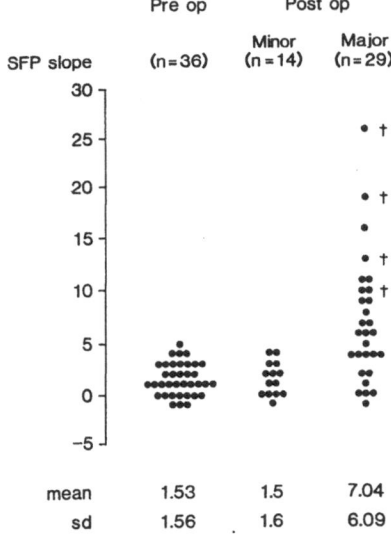

Fig. 4.2. The individual screen filtration pressure (SFP) slopes of 43 patients undergoing surgery are plotted. As the mean value + 2 × SD preoperatively and for patients having minor operations is 4.7, any SFP slope measurement above 5 was taken to indicate aggregates. Seventeen of the 29 patients undergoing major surgery had intravascular aggregates; the 4 patients who subsequently died within 10 days of surgery are marked with a cross.

SFP slope	Pre op (n=36)	Post op Minor (n=14)	Major (n=29)
mean	1.53	1.5	7.04
sd	1.56	1.6	6.09

Fig. 4.3a–d. On scanning electron microscopy in patients following minor surgery (**a**: ×200) there are few aggregates on the filter surface and the pores, which are 20 µm square, can clearly be seen. Following hemicolectomy this patient (**b**) had an SFP slope of 11 with over half the filter pores obscured by aggregates measuring up to 100 µm in diameter.

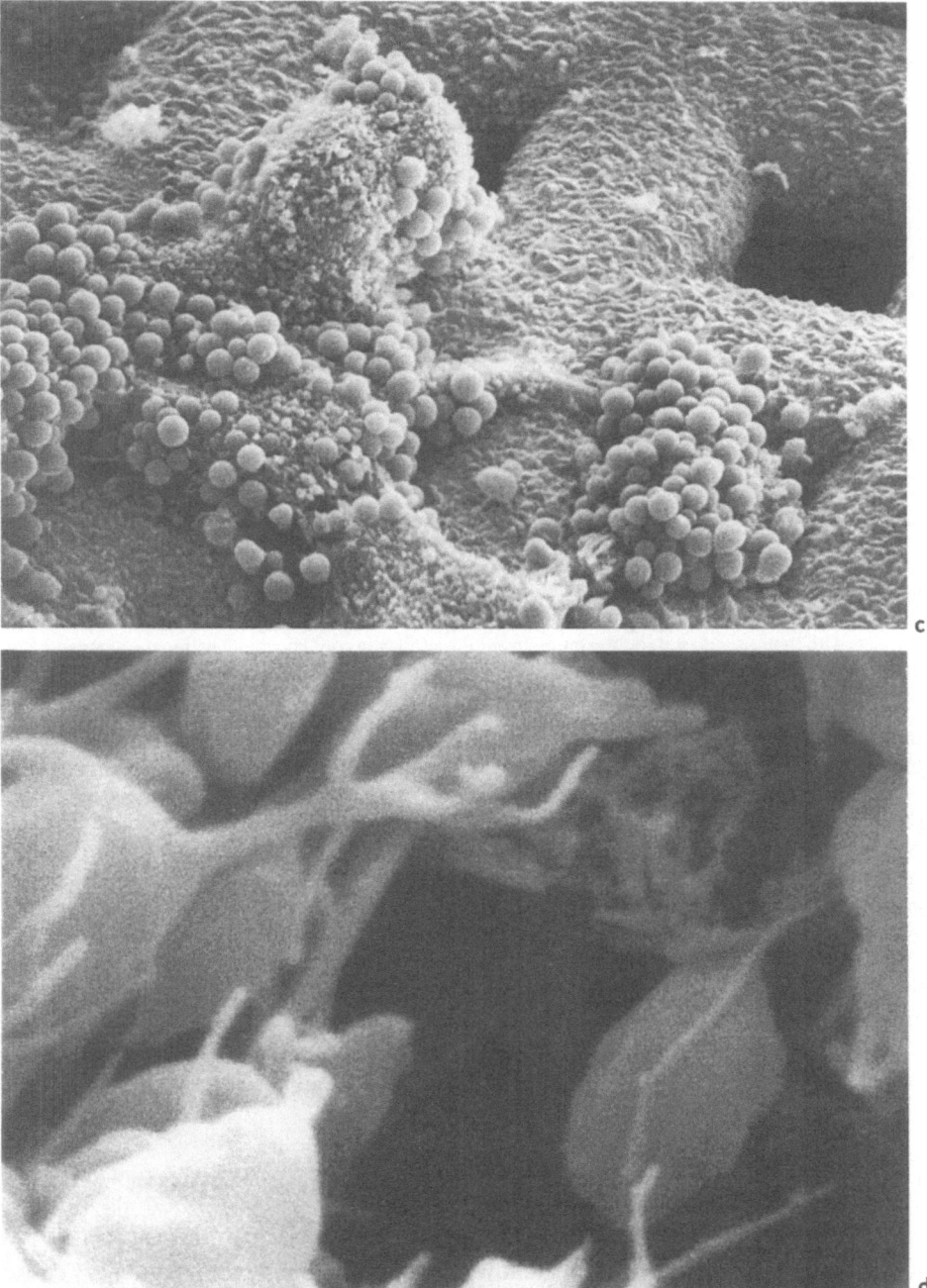

Fig. 4.3 *(continued).* At higher magnification (**c:** ×2000) the aggregate appears to consist of a mass of adherent and aggregated platelets. This is confirmed at high magnification (**d:** ×20 000) where highly activated platelets with pseudopod formation can be seen.

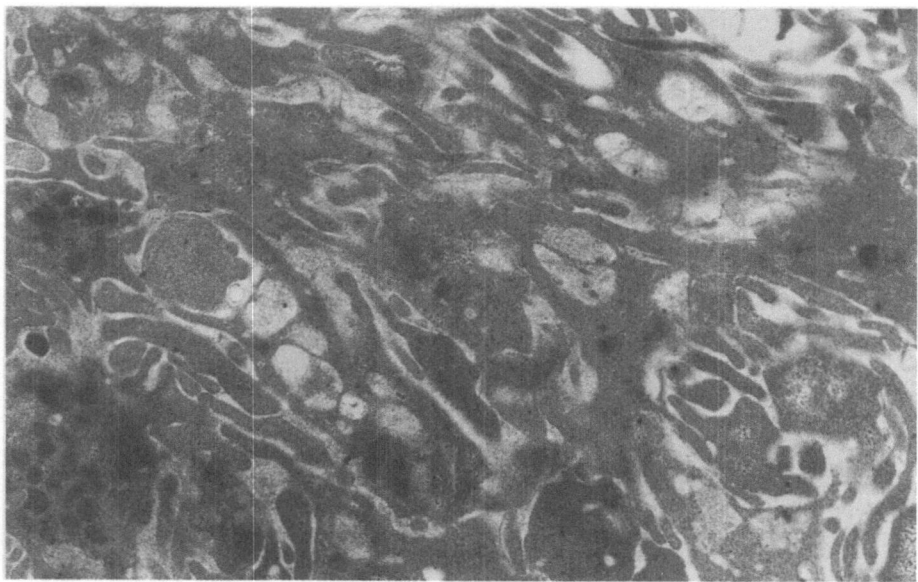

Fig. 4.4.   Transmission electron microscopy ($\times 15\ 000$) was performed to try and identify whether there was a leucocyte core. The aggregates uniformly consist of a mass of highly activated and degranulated platelets with their pseudopods tightly interwoven. Free collagen could be found within the aggregate but there was no core. The appearances are quite different from those of aggregates in stored blood.

deemed to be elevated and it is interesting that the 4 patients in the major surgery group who died within 10 days were among those with the highest SFP recordings.

The scanning electron micrographs of the filter surface in tests on blood from patients following major surgery show a mass of aggregates which appear to consist predominantly of platelets. Relatively clear filter surfaces are seen following minor surgery (Fig. 4.3). On transmission electron microscopy through the core of aggregates only densely packed and degranulating platelets are seen, with no evidence of activated leucocytes (Fig. 4.4). The appearance of those aggregates is entirely different from that of aggregates in stored blood, where non-viable cells loosely clump together without the dramatic pseudopod formation that is a feature of in vivo aggregates.

Intravascular Aggregates and Postoperative Hypoxia

Arterial hypoxia almost invariably follows abdominal surgery, with pulmonary complications classified as moderate or severe in nearly one third of patients (Greenhall et al. 1980). The usual explanations relating to general anaesthesia and painful wounds do not explain why pulmonary failure develops following major surgery but not following exploratory laparotomy and uncomplicated upper gastrointestinal surgery. As we have previously suggested postoperative pulmonary dysfunction consists of increased venous admixture with a rise in the alveolar–arterial difference in PO_2, which is a pattern indistinguishable from that described by Moore et al. (1969) for shock lung (Diament and Palmer 1967; Drummond 1975). It therefore seems possible that postoperative hypoxia may merely be a mild variety of shock lung with a similar aetiology. It does not seem reasonable to suppose that the pathophysiology that we call "shock lung" would always be so extreme that mortality was the usual outcome.

In our studies 20 patients underwent SFP measurement of aggregates in femoral vein blood 30 minutes and 3 hours following surgery; detailed evaluation of pulmonary function was performed preoperatively and on the first and seventh days following surgery. Patients with a SFP slope of 5 or greater in the postoperative period were allocated to the intravascular platelet aggregate-positive (IPA +ve) group and in these 10 patients the mean of the sum of SFP estimations at 30 minutes and 3 hours was 10.5 (range 6 to 24) compared with 2.4 (range −2 to 6) in the 10 IPA −ve patients.

Table 4.2. Pulmonary function: pre- and postoperative mean and standard error in 20 patients with (IPA +ve) and without (IPA −ve) detectable intravascular aggregates

	Arterial PO_2 (kPa)		Physiological dead space (ml)		Alveolar–arterial PO_2 gradient (kPa)	
	IPA −ve	IPA +ve	IPE −ve	IPA +ve	IPA −ve	IPA +ve
Preoperative	11.0 ±0.55	10.8 ±0.46	164.3 ±25.5	182.9 ±17.5	4.3 ±0.54	3.8 ±0.43
Postoperative						
Day 1	8.3[a] ±0.30	7.9[a] ±0.57	168.4 ±33.8	254.2 ±14.4	6.2[a] ±0.41	6.7[a] ±0.84
Day 7	10.4 ±0.47	9.0[b] ±0.53	152.4 ±19.0	193.4 ±22.0	4.7 ±0.31	6.1[b] ±0.56

[a] $p<0.05$ compared with preoperative values.
[b] $p<0.05$ compared with preoperative and with IPA −ve values.

The influence of the presence or absence of intravascular platelet aggregates on subsequent pulmonary function is shown in Table 4.2. There were no statistically significant differences in any of the pulmonary function parameters between the IPA +ve and IPA −ve groups either preoperatively or on the first postoperative day. On the seventh postoperative day, however, arterial oxygenation had recovered to near normal in the IPA −ve group but was still 1.8 ± 0.38 kPa lower than the preoperative value in IPA +ve patients ($p<0.05$) (Fig. 4.5). This change in arterial PO_2 was not due to any alteration in physiological dead space. It was, however, due to pulmonary arteriovenous shunting demonstrated by an increase in the alveolar–arterial oxygen gradient ($D(A–a)O_2$). This rose from 3.8 ± 0.43 kPa to 6.1 ± 0.56 kPa one week following surgery, a rise of 61% in the IPA +ve group. In those patients in whom there were no measurable intravascular platelet aggregates the $D(A–a)O_2$ recovered by the seventh day to a value similar to that preoperatively (Fig. 4.5b).

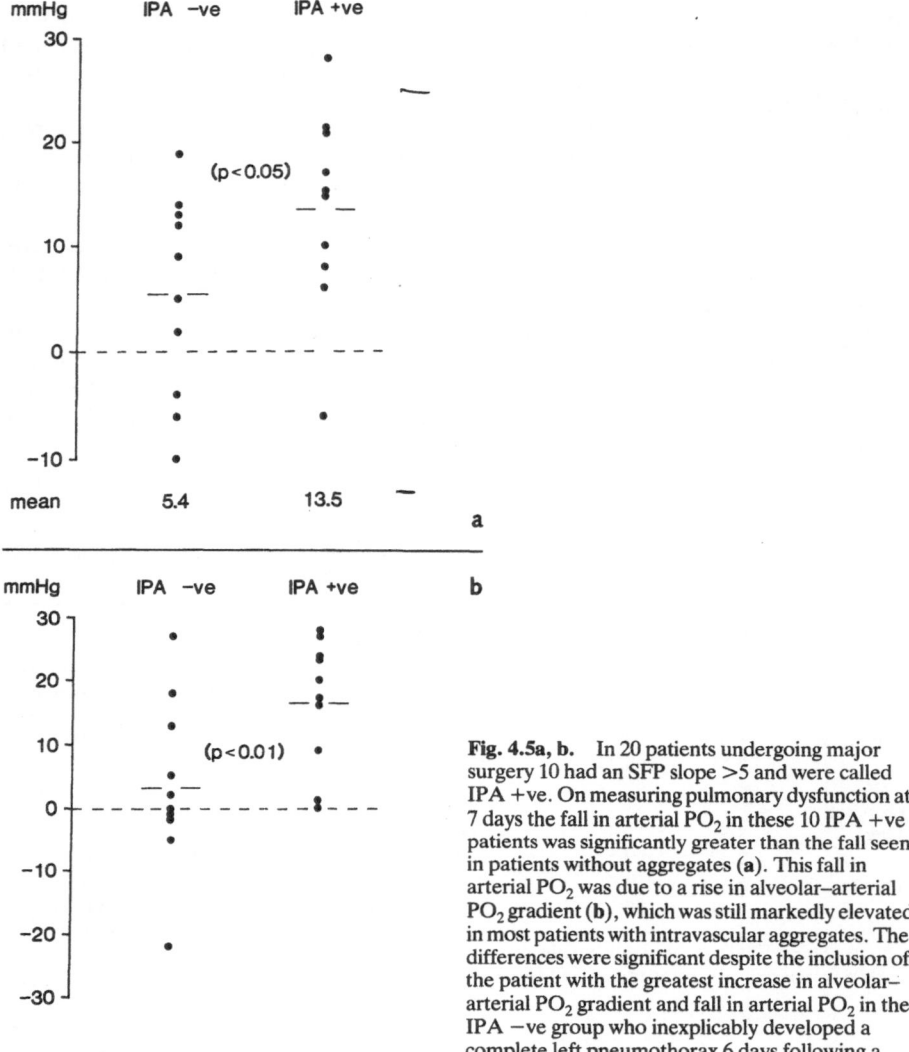

Fig. 4.5a, b. In 20 patients undergoing major surgery 10 had an SFP slope >5 and were called IPA +ve. On measuring pulmonary dysfunction at 7 days the fall in arterial PO_2 in these 10 IPA +ve patients was significantly greater than the fall seen in patients without aggregates (a). This fall in arterial PO_2 was due to a rise in alveolar–arterial PO_2 gradient (b), which was still markedly elevated in most patients with intravascular aggregates. The differences were significant despite the inclusion of the patient with the greatest increase in alveolar–arterial PO_2 gradient and fall in arterial PO_2 in the IPA −ve group who inexplicably developed a complete left pneumothorax 6 days following a gastrectomy.

Although many other potential risk factors for disturbed pulmonary function—including age, previous history of cardiopulmonary disease, cigarette smoking, volume of blood transfused and various haematological characteristics—were also measured the only other significant difference between the IPA −ve and IPA +ve patients was the fall in platelet count on the first postoperative day. This was 81.7×10^9 per litre in those patients with intravascular aggregates compared with only 21.4×10^9 per litre in the IPA −ve group ($p<0.02$). This fall in platelet count also significantly correlated with subsequent delayed postoperative hypoxia.

These results suggested that the early postoperative hypoxia that is so commonly seen after any laparotomy with general anaesthesia may well be due to the traditional explanations for postoperative hypoxia, which include painful wounds, mucus retention and segmental atelectasis of the lung due to critical closing volumes. We now believe that it is the delayed variety of postoperative pulmonary damage that equates most closely with shock lung. This may be particularly true as we know that shock lung develops at an interval following the precipitating episode. The possible mechanism for this will be discussed later.

The Influence of Platelet Kinetics

Before it can be argued that the intravascular platelet aggregates we measured were *the cause* of the postoperative pulmonary dysfunction with which they were associated it would be necessary to confirm that these aggregates accumulate in the lungs. The development of techniques for labelling platelets with isotopes has permitted a precise assessment of the movement of platelets to various vital organs during and following surgery or shock both in patients and in experimental models.

For our patient studies the platelets were labelled with ^{111}In-oxine by the method of Hawker et al. (1980) and reinjected on the day prior to aortic surgery in 20 consecutive patients. Platelet accumulation in the lung was measured by a scintillation counter positioned over the lung fields, the radioactivity being expressed as a ratio to that over the aortic arch as a blood pool. In this way the proportion of the radiolabelled platelets in the lung at any time could be calculated precisely. Furthermore the fall in circulating radioactivity was measured from 5 ml blood samples taken daily; this is a more precise measure of the disappearance of circulating platelets than the platelet count, which tends to measure aggregated platelets inaccurately. All measurements were taken on the day prior to surgery, 2½ hours postoperatively and then daily for 7 days. As with previous studies detailed pulmonary function was evaluated preoperatively and on the first and seventh days following surgery. The 20 patients who entered this study were also randomized to receive 25 mg kg^{-1} of methylprednisolone or an identical placebo at the beginning of their operative procedure. This dose was repeated if the procedure was not completed within 4 hours.

The pulmonary accumulation of radiolabelled platelets is plotted in Fig. 4.6. In the 24 hours prior to surgery there is no change in lung radioactivity but by 2½ hours postoperatively it has risen significantly. It is still higher the next day and remains elevated until the fourth postoperative day when it is no longer significantly greater than it had been in the preoperative period. Platelets therefore accumulate in the lungs following surgery but are removed within 3 or 4 days by a process that we presume involves fibrinolysis and phagocytosis. There was no difference between the methylprednisolone and the placebo group either in the pulmonary accumulation of platelets or in the screen filtration pressure measurement of femoral vein aggregates.

As all 20 patients had undergone aortic surgery it seemed appropriate to compare the pulmonary accumulation of platelets as measured by the rise in radioactivity (lung over aorta) at 24 hours following surgery with the subsequent fall in arterial PO_2 and rise in alveolar–arterial PO_2 gradient at 7 days (Fig. 4.7). The correlation coefficients were $r = 0.71$ and $r = 0.64$ respectively, which are highly significant and indicate that

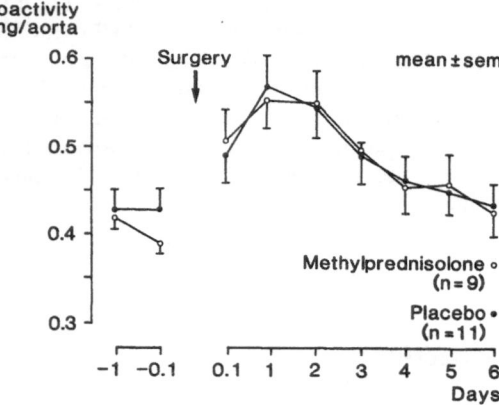

Fig. 4.6. In 20 patients randomized to methylprednisolone or placebo given intravenously at the start of aortic surgery there was rapid accumulation of radiolabelled platelets in the lung during the operation and in the first postoperative day. This platelet material, which had presumably embolized to the lungs, was cleared by 4 days.

in general a rapid accumulation of platelets within the lung over the period of surgical shock is associated with delayed pulmonary dysfunction one week later. The correlation between platelet kinetics and pulmonary dysfunction can be found in a simpler way: the fall in the platelet count over the operative 24 hours also correlates significantly with both the fall in arterial PO_2 and the rise in alveolar–arterial PO_2 gradient on the seventh postoperative day ($p<0.01$).

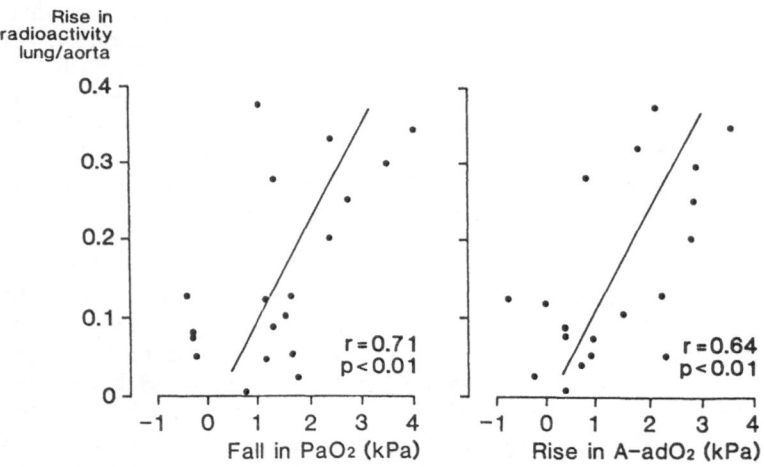

Fig. 4.7. The pulmonary accumulation of radiolabelled platelets measured as a rise in the ratio of radioactivity in the lung over that in the aorta is plotted against the fall in arterial PO_2 and the increase in alveolar–arterial PO_2 gradient on the seventh postoperative day. Of 20 patients undergoing aortic surgery those patients with the greatest accumulation of radiolabelled platelets in the lung in general had the worst pulmonary function, there being significant correlations with both a fall in arterial PO_2 and a rise in alveolar–arterial PO_2 gradient.

A Model for Surgical Shock

Despite all the growing evidence for the presence of microaggregates in peripheral blood in shock, for their accumulation in the lungs and for a positive relationship between pulmonary uptake of platelets and subsequent pulmonary dysfunction, this relationship could still be coincidental. It would be reasonable to argue that severe shock causes both the greatest fall in platelet count and the most severe pulmonary dysfunction by mechanisms independent of each other. In order to prove a causal relationship it would be necessary to demonstrate that the prevention of intravascular aggregate formation by prostaglandin manipulation or some other means resulted in a reduction in the severity of subsequent pulmonary dysfunction. It is obviously quite inappropriate to evaluate new agents with the potential of preventing aggregate formation in patients, and for this reason an animal model of the clinical situation would seem to be essential.

As the pig has similar haematology to humans this species was selected. Our experience with shock is greatest as regards major surgery and for this reason our model was one that mimics the clinical conditions during aortic replacement (Fig. 4.8). As in our patient studies the pig's platelets were labelled with ^{111}In-oxine 24 hours prior to surgery. Under general anaesthesia carotid arterial and Swan–Ganz pulmonary artery catheters were placed allowing measurement of arterial blood pressure, central venous pressure, pulmonary artery and pulmonary wedge pressure, cardiac output, and peripheral and pulmonary vascular resistance. Surgery consisted of a full midline laparotomy with the small bowel exposed for 30 minutes and the

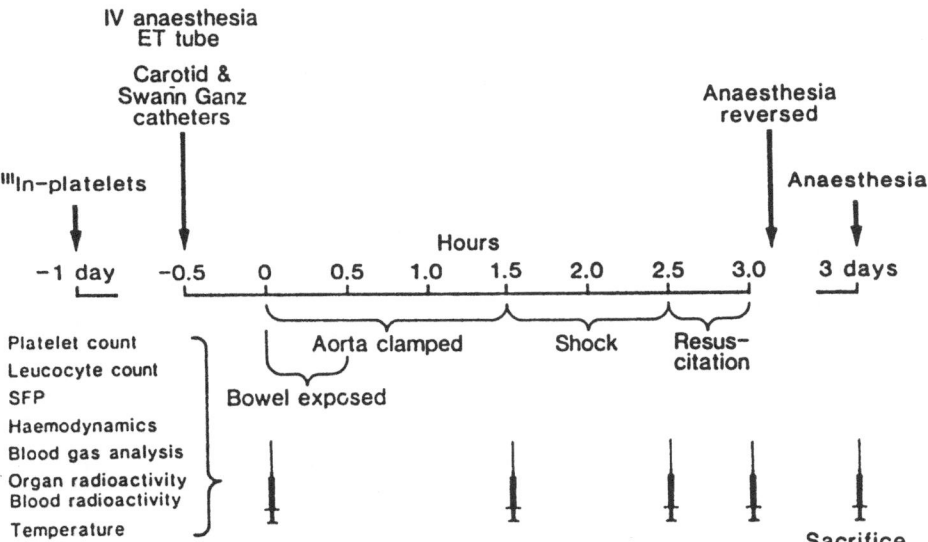

Fig. 4.8. In our model for major surgical shock pigs whose autologous platelets have been labelled with ^{111}In, undergo laparotomy with small bowel exposure and 1.5 hours of aortic clamping. Following removal of the aortic clamp shock is allowed to continue for 1 hour prior to resuscitation. Extensive haemodynamic, platelet kinetic and pulmonary function measurements are made at the intervals shown.

infrarenal aorta cross-clamped for 1.5 hours. During this period there was a gradual
development of tachycardia with hypotension but shock became more severe on
removal of the aortic cross-clamp. This was allowed to continue for a further hour
before vigorous resuscitation with Haemaccel and clear fluids which was performed
over half an hour, prior to reversal of anaesthesia. The animal was then carefully
monitored over the next 3 days, at which time it was reanaesthetized, all measurements repeated and then killed for excision of the vital organs and measurement of
their radioactivity. Other measurements included platelet and leucocyte counts, SFP
measurement of circulating aggregates, blood gas analysis, radioactivity of the liver,
spleen, kidneys and lungs, and plasma opsonic function; these were all taken prior to
surgical shock, just before removal of the aortic cross-clamp, following the episode
of severe shock, following resuscitation and then at 3 days prior to the animal being
killed.

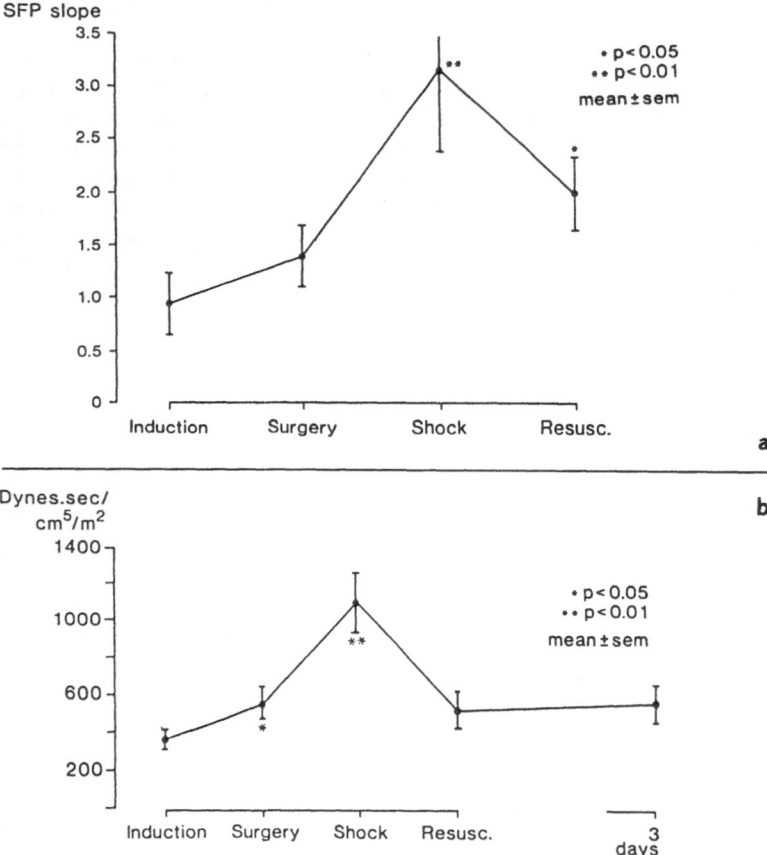

Fig. 4.9a–d. During surgical shock in the pig model there is a clear sequence of events. Shortly after the
start of surgery SFP of inferior vena caval blood rises indicating the presence of intravascular aggregates
(**a**). These reach a peak following the removal of the aortic clamp during shock and at this time pulmonary
vascular resistance rises dramatically (**b**). There is little pulmonary arteriovenous shunting during surgery
and shock but when the cardiac output rises following resuscitation arteriovenous shunts open (**c**). Predictably the changes in alveolar–arterial oxygen difference (**d**) follow the changes seen in pulmonary
arteriovenous shunting.

This model reliably produced shock in the pig, with the mean ($\pm$ SEM) arterial pressure falling from 116 ± 6 mmHg at induction to 70 ± 6 mmHg during the shock period ($p<0.01$). During the same interval cardiac output fell from $2.3 \pm 0.2 \, l \, min^{-1}$ to $1 \pm 0.1 \, l \, min^{-1}$ during surgical shock, although it recovered to normal following resuscitation ($p<0.01$). Throughout the study period both platelet and leucocyte counts fell, with the platelet count falling from $437 \pm 48 \times 10^9 \, l^{-1}$ to $252 \pm 39 \times 10^9 \, l^{-1}$ following resuscitation and further to $98 \pm 13 \times 10^9 \, l^{-1}$ at 3 days ($p<0.01$). These changes in platelet count were reflected in similar changes in the blood radioactivity, indicating the rapid consumption of platelets. During surgery and shock the SFP slope measuring intravascular aggregates rose from 1.0 ± 0.3 to 3.2 ± 0.8 during the shock episode ($p<0.01$) (Fig. 4.9). During the same time interval radiolabelled platelet accumulation in the lungs increased pulmonary radioactivity progressively; a peak was reached during surgical shock but values remained elevated even at 3 days following surgery ($p<0.01$) (Fig. 4.10). This accumulation of platelets within the lungs was associated with a rise in pulmonary vascular resistance, with the resistance index increasing from 364 ± 45 to a peak of 1096 ± 165 during surgical shock (Fig. 4.9).

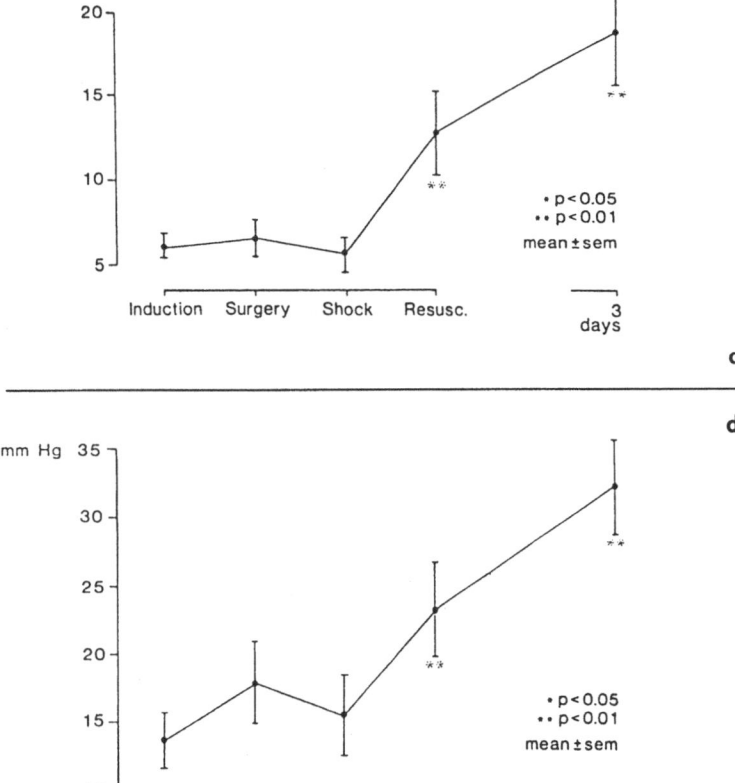

Fig. 4.9 (continued)

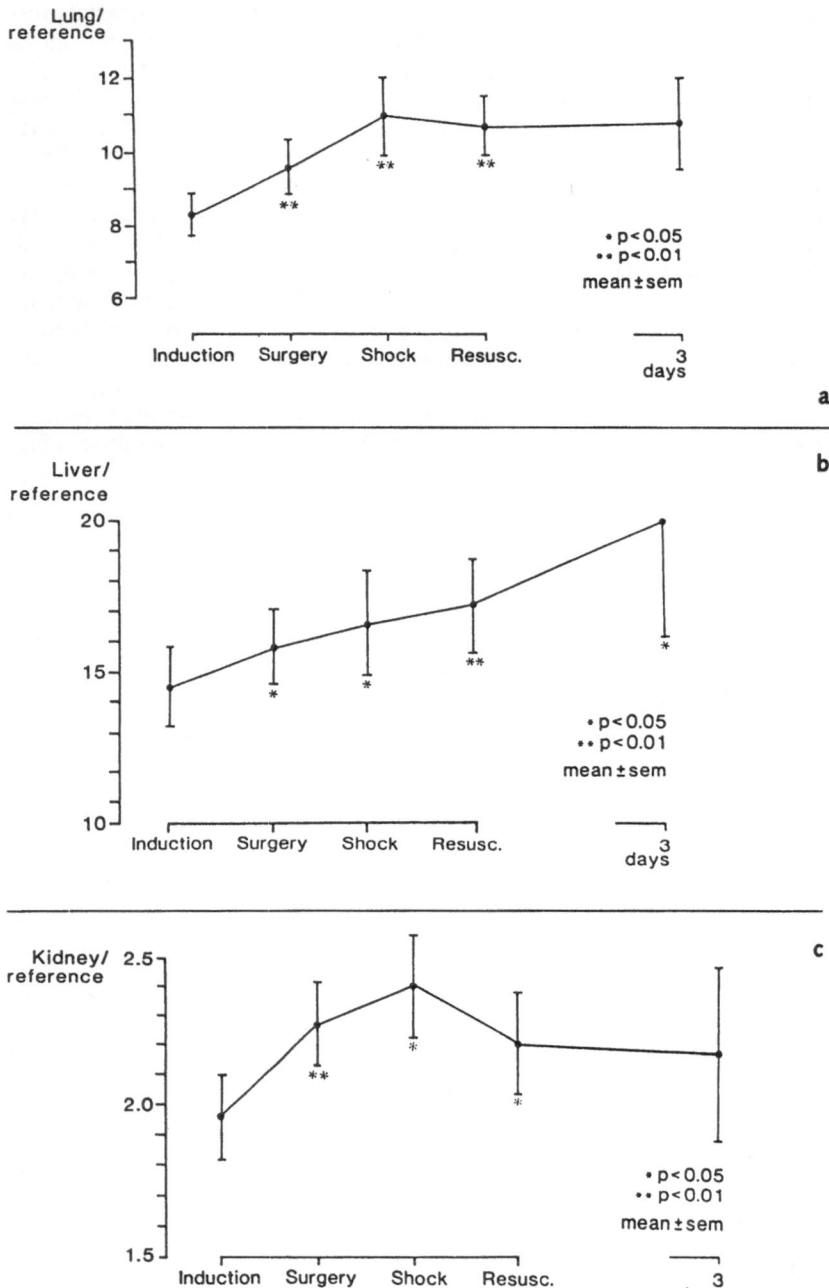

Fig. 4.10a–c. In the pig shock model the radioactivity of the vital organs was compared with that in the aortic arch as a blood pool, using a highly collimated probe. Throughout the study period there was a significant rise in the radioactivity of the vital organs suggesting platelet microembolization. At 3 days only 7 of the pigs had survived and hence statistical significance was lost for the rise in radioactivity observed in the lung and kidney.

Following resuscitation the pulmonary vascular resistance fell but measurements of pulmonary arteriovenous shunting indicated that this was substantially due to shunt flow, with the pulmonary shunt rising from $5.8 \pm 1.0\%$ during shock to $12.9 \pm 2.4\%$ following resuscitation and still higher to $18.9 \pm 3.5\%$ at 3 days ($p<0.01$) (Fig. 4.9). Predictably this was reflected in a rapid rise in the alveolar–arterial oxygen difference which reached an impressive 32.5 ± 3.4 mmHg at 3 days postoperation (Fig. 4.9). This produced a fall in arterial oxygen tension from 104 ± 6 mmHg to 70 ± 4 mmHg at 3 days in those animals that survived to this interval ($p<0.01$). Lung histology demonstrated pulmonary atelectasis, oedema and congestion in all animals, although this was most severe in those animals that died in the postoperative period.

This experimental model has given us a unique opportunity to study the mechanisms of pulmonary damage during surgical shock. There is a clearly defined sequence of events, with shock stimulating a fall in platelet and leucocyte counts, and a rise in circulating aggregates accompanied by a rise in pulmonary vascular resistance. During the shock episode cardiac output is low but on resuscitation cardiac output increases and the pulmonary circulation accommodates for this by increasing shunt flow. The increase in pulmonary shunt is associated with an increase in the alveolar–arterial PO_2 gradient and arterial hypoxaemia results. Radiolabelled platelet studies demonstrate that at the same time as pulmonary vascular resistance is rising platelet accumulation is maximal in the lungs (Fig. 4.10). Finally, to emphasize that the other vital organs may be damaged by a similar process we find that there is also a progressive uptake of platelets within the liver and kidney with radioactivities rising during surgical shock from 14.5 ± 1.3 to 17.3 ± 1.6 ($p<0.01$) and 2.0 ± 0.1 to 2.4 ± 0.2 ($p<0.05$) respectively. This suggests that a similar mechanism may be responsible for the deterioration in liver and renal function that is known to occur following surgery and which may contribute to mortality in shocked patients receiving intensive care.

Hypothesis on the Pathophysiology of Shock Lung

In considering the mechanism of lung damage following shock the delay between the shock episode and the development of severe pulmonary dysfunction has largely been ignored and yet is critical to the underlying pathophysiology. From our studies we have evidence for the formation of intravascular aggregates and for their accumulation in the lungs. How do they promote delayed pulmonary arteriovenous admixture?

Aggregate size on scanning electron microscopy varies from less than 10 µm to greater than 200 µm (Fig. 4.3). These aggregates might be distributed randomly with pulmonary blood flow, with most perhaps being carried to the well-perfused areas of the lung. Smaller aggregates would lodge in alveolar arterioles, which vary from 5 µm to 30 µm in diameter (Krahl 1965). The obstructed capillary bed may then in part thrombose with the release of kinins, serotonin, histamine and other inflammatory substances from both the aggregate and the secondary thrombosis (Fig. 4.11). At this stage a number of alveolar units within the lung would be damaged with an

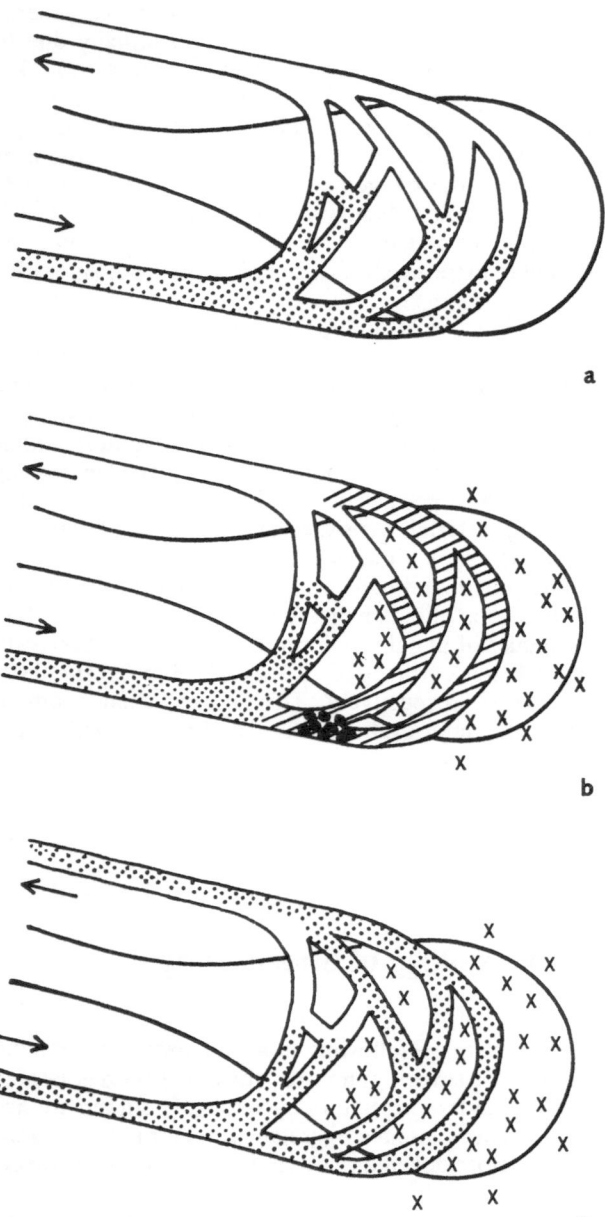

Fig. 4.11a–c. A possible pathophysiological mechanism for shock lung is explained in these diagrams representing a single alveolus (**a**). An aggregate embolizing to the alveolar arterioles will promote secondary thrombosis and through the release of kinins an inflammatory reaction (**b**). Pulmonary shunting does not, however, develop until the aggregate and the secondary thrombosis are removed by phagocytosis or the extensive fibrinolytic activity of the lung. Once this occurs deoxygenated blood from the pulmonary artery may pass through the damaged alveolar circulation to produce pulmonary arteriovenous admixture (**c**).

increase in dead space as these are poorly perfused and yet still ventilated alveoli. In this situation hypoxaemia would only occur should large numbers of aggregates damage a substantial proportion of alveoli, as oxygenation can be maintained by hyperventilation. Over the following days the aggregate and the secondary thrombosis will be phagocytosed or lysed by the rich fibrinolytic capacity of the lung vasculature. The resulting reperfusion of a damaged alveolus is depicted in Fig. 4.11(c) and represents the conditions in which venous admixture develops. Venous blood in the now recannalized circulation passes through the still-damaged alveolus with a diffusion block secondary to the inflammatory process of thrombosis.

Where larger aggregates obstruct terminal pulmonary arteries of diameter 25 μm up to 300 μm (Krahl 1965) a more serious lesion may result: generalized pulmonary microembolization impairs collateral circulation so that with collapse of the involved alveoli ischaemic alveolar damage may result (Modry et al. 1977). Patches of infarction may develop producing parenchymatous haemorrhage, more severe oedema and fibrosis over a period of weeks (Fig. 4.12). Again as the vessels are recannalized pulmonary shunting and venous admixture will ensue. On this basis it is possible to explain the various findings in previous studies on shock lung and the failure to find aggregates within the lungs of patients who die more than 48 hours following injury.

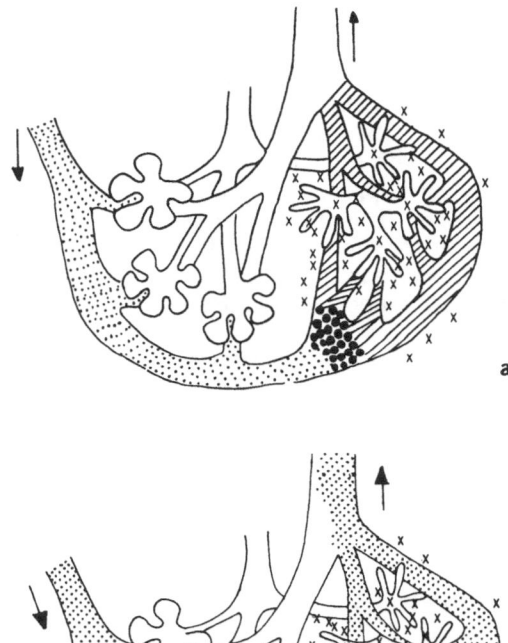

Fig. 4.12a, b. When an embolus >30 μm embolizes it may obstruct the pulmonary arterial supply to several alveoli (**a**). As a result these alveoli may collapse and if the collateral circulation is also damaged by aggregates ischaemic damage may result. On phagocytosis and lysis of the aggregate and thrombus a more severe and possibly protracted venous admixture will result.

The Reticuloendothelial System

Pulmonary accumulation of aggregates clearly reflects a failure of the reticulo-endothelial system to clear particulate matter from the blood. In health the liver, spleen and lungs are capable of processing substantial quantities of debris without visible damage, but in shock the entire system is suppressed and may easily be over-whelmed (Pardy and Dudley 1977). Although the mechanism of reticuloendothelial suppression is not yet clear, we believe that the consumption of plasma opsonic factors may be important. These substances, which include fibronectin, coat the surface of particulate debris so that it may be more easily engulfed within the reticuloendothelial system. In both endotoxic and surgical shock we have found that fibronectin is grossly depleted and the remaining level of fibronectin is functionally deficient as it exists predominantly in complexes containing collagen debris (Powell et al. 1986).

Possible Methods of Preventing Aggregate Formation

The hypothesis outlined in this chapter can only be verified by preventing aggregate formation with resulting improvement in pulmonary function following major surgery or shock. The therapeutic options include platelet inhibitory drugs, of which those substances with prostaglandin activity would seem most promising. Such substances may be examined firstly in a rabbit endotoxic model and then in our pig major surgery study. A viable alternative for preventing hypoxia following major surgery would be the selective removal of platelets, leucocytes and clotting factors by continuous flow centrifugation, allowing their replacement at the end of the surgical procedure. Clearly this approach would have little or no value in the treatment of patients with septicaemia or following trauma. We believe that it is important to study such approaches in a reproducible model where the shock may be precisely controlled. The objectives must be to find methods of prophylaxis that reduce both the incidence and the severity of this pulmonary condition which has, over the last 20 years, been a frequent cause of death in often young and salvageable patients recovering from shock and major surgery.

We are deeply grateful to the Upjohn Corporation for their continued support of this project over the last six years. Keith Poskitt was funded through the generosity of the Special Trustees of Charing Cross Hospital and we would like to thank the British Heart Foundation for their contribution. We are grateful for the technical expertise of Joseph Irwin and for the experience in pulmonary function measurements and anaesthetic support of Dr Wolfgang Kox.

References

Attar S, Mansberger AR, Irani B, Kirby W, Masaitis C, Cowley RA (1966) Coagulation changes in clinical shock. II. Effect of septic shock on clotting times and fibrinogen in humans. Ann Surg 164:41–50

Bailey ME (1976) Endotoxin, bile salts and renal function in obstructive jaundice. Br J Surg 63:774–778

Bergofsky EH (1970) The adult acute respiratory insufficiency syndrome following non-thoracic trauma: the lung in shock. Am J Cardiol 26:619–621

Blaisdell FW, Lewis FR (1977) Respiratory distress syndrome of shock and trauma. WB Saunders, Philadelphia

Blaisdell FW, Lim RC, Stallone RJ (1970) The mechanism of pulmonary damage following traumatic shock. Surg Gynecol Obstet 130:15–22

Bredenberg CE, Taylor GA, Webb WR (1980) The effect of thrombocytopenia on the pulmonary and systemic hemodynamics in canine endotoxin shock. Surgery 87:59–68

Brown C, Dhurandhar HN, Barrett J, Litwin MS (1977) Progression and resolution of changes in pulmonary function and structure due to pulmonary microembolism and blood transfusion. Ann Surg 185:92–99

Clowes GHA, Fabbington GH, Zuschneid W, Cossette GR, Saravin C (1970) Circulating factors in the etiology of pulmonary insufficiency and right heart failure accompanying severe sepsis (peritonitis). Ann Surg 171:663–678

Collins JA (1969) The cause of progressive pulmonary insufficiency in surgical patients. J Surg Res 9:685–704

Collins JA, James PM, Bredenberg CE, Anderson RW, Heisterkamp CA, Simmons RL (1978) The relationship between transfusion and hypoxaemia in combat casualties. Ann Surg 188:513–520

Cook WA, Webb WR (1968) Pulmonary changes in haemorrhagic shock. Surgery 64:85–94

Craddock PR, Fehr J, Brigham KL, Kronenberg RS, Jacob HS (1977) Complement and leukocyte-mediated pulmonary dysfunction in haemodialysis. N Engl J Med 296:769–774

Del Guercia LR, Cohn JD, Greenspan M, Feins NR, Kornitzer G (1966) Pulmonary and systemic arteriovenous shunting in clinical septic shock. In: Brown IW, Cox BG (eds) International conference on hyperbaric medicine. National Academic Science, Durham NC

Diament ML, Palmer KNV (1967) Venous/arterial pulmonary shunting as the principal cause of post-operative hypoxaemia. Lancet I:15–17

Drummond GB (1975) Postoperative hypoxaemia and oxygen therapy. Br J Anaesth 47:211–228

Fuchsig F, Brucke P, Blumel G, Gottlob R (1967) A new clinical and experimental concept of fat embolism. N Engl J Med 276:1192–1193

Geelhoed GW, Bennett SH (1975) "Shock lung" resulting from perfusion of canine lungs with stored bank blood. Am Surg 41:661–682

Greenhall MJ, Evans M, Pollock AV (1980) Midline or transverse laparotomy? A random controlled clinical trial. II. Influence on postoperative pulmonary complications. Br J Surg 67:191–194

Hammerschmidt DE, Weaver LJ, Hudson LD, Craddock PR, Jacob HS (1980) Association of complement activation and elevated plasma-C5a with adult respiratory distress syndrome: pathophysiological relevance and possible prognostic value. Lancet I:947–949

Hardaway RM (1962) The role of intravascular clotting in the aetiology of shock. Ann Surg 155:325–338

Hardaway RM, Husni EA, Geever EF, Noyes HE, Burns JW (1961) Endotoxin shock: a manifestation of intravascular coagulation. Ann Surg 154:791–802

Harrison MW, Connel RS, Campbell JR, Webb MC (1977) Microcirculatory changes in the lung of the hypoxic and hypovolaemic puppy. Ann Surg 185:311–317

Hawker RJ, Hawker LM, Wilkinson AR (1980) Indium (III-In)-labelled human platelets: optimal method. Clin Sci 58:243–248

Hewson W (1772) In: The works of William Hewson FRS, ed G Gulliver. Sydenham Society, London (1846)

Hissen W, Swank RL (1965) Screen filtration pressure and pulmonary hypertension. Am J Physiol 209:715–722

Jacob HS (1978) Granulocyte–complement interaction. A beneficial antimicrobial mechanism that can cause disease. Arch Intern Med 138:461–463

Krahl VE (1965) The lung as a target organ in thromboembolism. In: Sasahara AA (ed) Pulmonary embolic disease. Grune and Stratton, New York

Lillehei RC, Longerbeam JK, Bloch JH, Manax WG (1964) The nature of irreversible shock. Ann Surg 160:682–710

Lim RC, Blaisdell FW, Goodman JR, Hall AD, Thomas AN (1966) Electron microscopic study of pulmonary microemboli in regional and systemic shock. Surg Forum 17:25–27

McCollum CN, Campbell IT (1979) Value of measuring intravascular platelet aggregates in the prediction of post-operative pulmonary dysfunction. Br J Surg 66:703–707

McNamara JJ, Burran EL, Larson E, Omiya G, Suehire G, Yamase H (1972) Effect of debris in stored blood on pulmonary microvasculature. Ann Thorac Surg 14:133–139

Modry DL, Chiu CJ, Jinchey EJ (1977) The roles of ventilation and perfusion in lung metabolism. J Thorac Cardiovasc Surg 74:275–285

Moore FD, Lyons JH, Pierce EC, Morgan AP, Drinker PA, MacArthur JD, Dammin GJ (1969) Post-traumatic pulmonary insufficiency. WB Saunders, Philadelphia

Moseley RV, Doty DB (1970) Death associated with multiple pulmonary emboli soon after battle injuries. Ann Surg 171:336–346

Murray JF (1977) Mechanisms of acute respiratory failure. Am Rev Respir Dis 115:1071–1078

Nahas RA, Melrose DG, Sykes KM, Robinson B (1965) Post-perfusion lung syndrome: role of circulatory exclusion. Lancet II:251–254

Nash G, Bowen JA, Langlinais PC (1971) Respirator lung, a misnomer. Arch Pathol 21:234–240

Pardy BJ, Dudley HAF (1977) Post-traumatic pulmonary insufficiency. Surg Gynecol Obstet 144:259–269

Powell JP, Poskitt KR, Irwin JTC, Attanoos RL, McCollum CN (1986) Opsonic dysfunction secondary to plasma fibronectin depletion after aortic surgery. Br J Surg 73:38–40

Robb HJ (1963) The role of microembolism in the production of irreversible shock. Ann Surg 175:685–697

Shah-Mirany J, Najafi H, Serry C, Callaghan R, Yang J (1970) Pathophysiological alterations in perfused and non-perfused lungs during cardiopulmonary bypass. Ann Thorac Surg 10:402–408

Simmons RL, Ducker TD, Anderson RW (1968) Pathogenesis of pulmonary oedema following head trauma. J Trauma 8:800–811

Stein M, Thomas DP (1967) Role of platelets in the acute pulmonary response to endotoxin. J Appl Physiol 23:47–52

Swank RL (1961) Alteration of blood on storage: measurement of adhesiveness of aging platelets and leukocytes and their removal by filtration. N Engl J Med 265:728–733

Tiefenbrun J, Dikman S, Shoemaker WC (1975) The correlation of sequential changes in the distribution of pulmonary blood flow in haemorrhagic shock with the histopathologic anatomy. Surgery 78:618–627

Ygge J (1970) Changes in blood coagulation and fibrinolysis during the postoperative period. Am J Surg 119:225–232

Section II

Pathophysiology of Septic Shock and the Adult Respiratory Distress Syndrome

Chapter 5

Biochemical Changes in Patients at Risk from the Adult Respiratory Distress Syndrome: Does the Pancreas Play a Role?

M. Lamy, M. E. Faymonville, A. Adam, G. Deby-Dupont,
L. Bodson, P. Damas and P. Franchimont

The adult respiratory distress syndrome (ARDS) is the clinical manifestation of a pathophysiological process of rapid onset, characterized by severe dyspnoea, hypoxaemia, increased lung stiffness and diffuse bilateral pulmonary infiltrates. It results from direct or non-specific pulmonary damage to previously normal lungs. The mechanism of acute lung injury in ARDS remains unknown, but different mediator systems are likely to play a role in the pathogenesis of this syndrome. Two recent papers (Deby-Dupont et al. 1984; Nicod et al. 1985) have pointed to a potential role for pancreatic enzymes in the pathogenesis of ARDS, particularly in patients with sepsis, where highly abnormal plasma levels of immunoreactive trypsin have been found. When liberated into the systemic circulation trypsin is a potent protease which can induce damage in very different organs, as observed in acute pancreatitis. Furthermore, trypsin may activate all the other proenzymes including the kallikrein–kininogen–kinin system (KKK system). This may lead to the generation of kinins, especially bradykinin, one of the supposed mediators of ARDS (Fig. 5.1).

In order to evaluate the potential role of pancreatic enzymes and the role of the KKK system we investigated 20 critically ill patients admitted to our Intensive Care Unit (ICU) who were at risk of ARDS. We measured plasma concentrations of pancreatic enzymes, antiproteases, prokallikrein and kininogens on a daily basis. We also measured the acute phase proteins which, like prokallikrein and kininogens, are synthesized by the liver, but are not activated at the site of the injury (Adam et al. 1985).

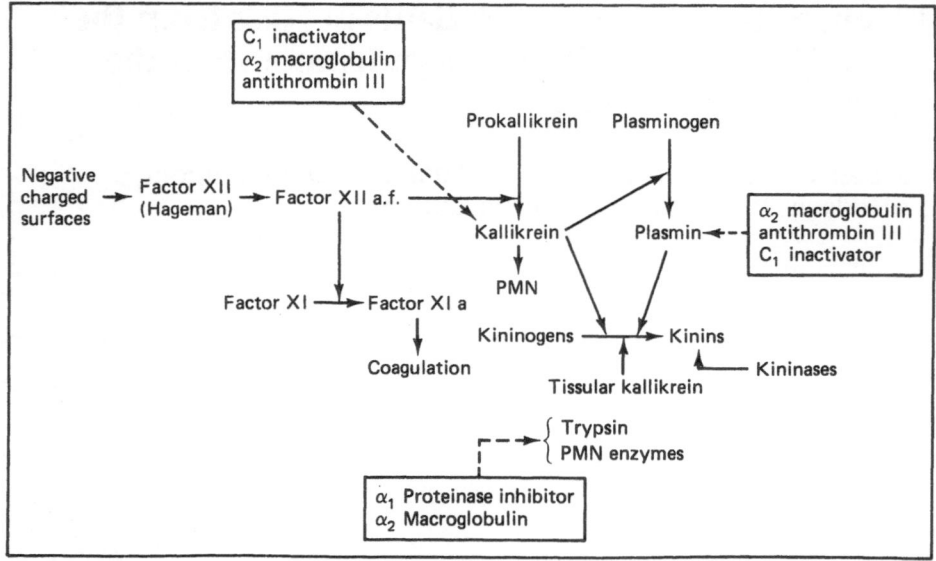

Fig. 5.1. Schematic representation of the kallikrein–kininogen–kinin system, its relation to fibrinolysis and the initial stage of the intrinsic clotting system. Inhibitors are represented by *interrupted arrows*.

Materials and Methods

Patient Groups and ARDS Diagnostic Criteria

Twenty patients (mean age 51.8 ± 13 (range 26–82); 13 male, 7 female) who had been admitted to the ICU of the University Hospital of Liège and were at risk of developing ARDS, were selected. Their diagnosis and final outcome are shown in Table 5.1. The abdominal surgery patients had undergone major surgery and

Table 5.1. Description and final outcome of the patients

Diagnosis	n			Survival	
	Total	ARDS	non-ARDS	In ARDS group	In non-ARDS group
Abdominal surgery	12	5	7	1	6
Multiple injury	6	2	4	2	4
Drowning	2	1	1	1	1
Total	20	8	12	4 (50%)	11 (91.6%)

required prolonged intensive care postoperatively; 7 patients had had an emergency abdominal aortic replacement with massive blood transfusion, while 5 had undergone gastrointestinal surgery with severe septic complication. The multiple injury patients had at least three major injuries (head, chest, abdomen or limbs and pelvis). Two patients had had a drowning accident with massive fresh water inhalation.

ARDS was defined as present if the patient developed non-cardiogenic pulmonary oedema (radiological opacification of both lungs with no increase in pulmonary capillary wedge pressure) and severe arterial hypoxaemia requiring artificial ventilation with an inspired fraction of oxygen higher than 0.5, a tidal volume of 12–15 ml kg^{-1} body weight and a positive end-expiratory pressure of at least 5 cm H_2O in order to normalize gas exchange. These criteria correspond to the groups of moderate or severe acute respiratory failure according to the scoring system of the Massachusetts General Hospital (Pontoppidan et al. 1985).

Blood Sampling

Blood samples were collected in the 20 patients within 24 hours of admission to the ICU; further sampling was performed twice daily during the first 2 days of admission, and then once daily for up to a maximum of 12 days until the patient either left the ICU or died. Blood was gently drawn from short catheters into polystyrene tubes containing sodium citrate (0.1 mmol, 1/10 v/v) for measurements of kininogens and prokallikrein, and into heparin for trypsin determination. Blood was allowed to clot and serum was used for determination of amylase, lipase, α_1-proteinase inhibitor (α_1PI), α_2-macroglobulin (α_2M), antithrombin III (AT III), C1 esterase inhibitor (C1-inh) and acute phase proteins (C-reactive protein, α_1 acid glycoprotein and haptoglobin). All blood samples were immediately centrifuged and plasma or serum was stored at $-30°C$ until assayed. Quantification was performed within a month.

Immunoreactive trypsin (IRT: free trypsin, trypsinogen and the complex trypsin–α_1PI) was measured by radioimmunoassay using a rabbit antiserum to human cathodic trypsin, ^{125}I-labelled human cathodic trypsin, and non-labelled cathodic trypsin standards (Geokas et al. 1979). The lower limit of sensitivity was 5 µg l^{-1} plasma (but using this immunological technique the complex trypsin–α_2M cannot be detected). Enzymic methods were used for lipase and amylase determinations. The determinations of α_1PI, α_2M, AT III, C1-inh and acute phase proteins were by laser nephelometry. Total, low molecular weight and high molecular weight kininogens were determined by specific radioimmunoassay according to the methods described by Adam et al. (1985a).

Prokallikrein enzymic determination has also been well described in detail by Adam et al. (1985b).

Statistical Analysis

Means and standard deviations were calculated using a TRS 80 computer. An analysis of variance was used for comparison of two groups (ARDS and non-ARDS).

Results

The clinical data of the 20 patients are summarized in Table 5.1; 8 of them developed ARDS, the majority of the patients being in the abdominal surgery group. ARDS occurred on day 0 after drowning; on days 1 to 2 after multiple injury and abdominal vascular surgery; and on days 4 and 5 following gastrointestinal surgery. Mortality was higher in patients who developed ARDS compared with those who did not. This is in accordance with the results of most other studies. A comparison between these two groups, ARDS and non-ARDS, was made for all the measured biochemical parameters.

Fig. 5.2 shows changes in plasma concentration of IRT (ng ml^{-1}), amylase (IU l^{-1}) and lipase (IU l^{-1}) in the two patient groups from day 0 to day 12. Abnormal IRT values were found in both the ARDS and the non-ARDS groups from day 1 until the end of the observation period. No statistically significant difference between the two groups was observed. Mean plasma activity of amylase was abnormal for the first 5 days after injury in the ARDS group, whereas in the non-ARDS group higher amylase values were noted later (days 5 to 10). Lipase mean plasma activity in the

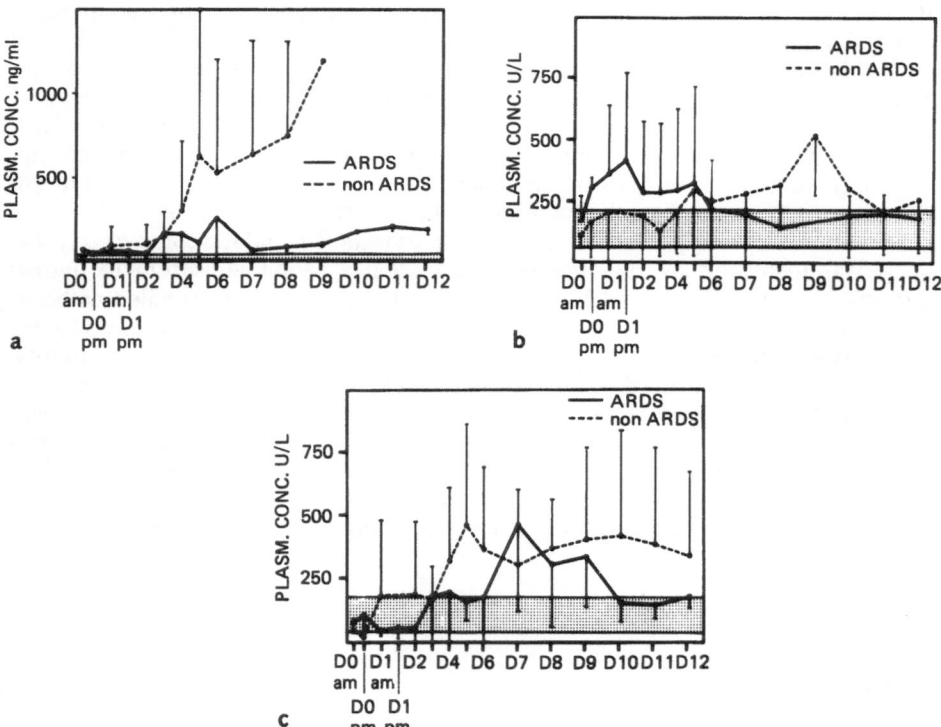

Fig. 5.2a–c. Changes in plasma concentrations of immunoreactive trypsin (**a**), amylase (**b**) and lipase (**c**) in 20 critically ill patients over 12 days. The *stippled areas* represent the normal range of values in healthy volunteers. D0 = day of injury or surgery.

non-ARDS group began to rise on day 3 and was abnormal on day 4. However, it only became abnormal in the ARDS group on day 7.

Fig. 5.3 shows the evolution of changes in plasma concentration of four proteinase inhibitors. Mean α_1PI plasma concentration was low but in the normal range on day 0; it increased on day 1 in the two groups and values were greater than normal on day 3. In the ARDS patient group, plasma concentration increased further up to day 12. In non-ARDS patients mean values remained just above the normal range. A statistically significant difference appeared between the two groups on days 11 and 12. As also shown in Fig. 5.3, mean α_2M plasma concentrations were below normal and decreased in ARDS patients from day 0 up to day 12. In non-ARDS patients the mean value of α_2M on day 0 was in the normal range but decreased during the observation period; however, these levels were always higher than in patients with ARDS. Plasma concentration changes of AT III were similar in the two patient groups. The levels remained at the lower limit of the normal range, and abnormally low levels were observed in the ARDS group on day 0 p.m. and day 1 p.m. (Fig. 5.3). C1-inh plasma levels were normal in the two groups early after injury (days 0 and 1) but afterwards increased, with no significant differences being seen between ARDS and non-ARDS patients.

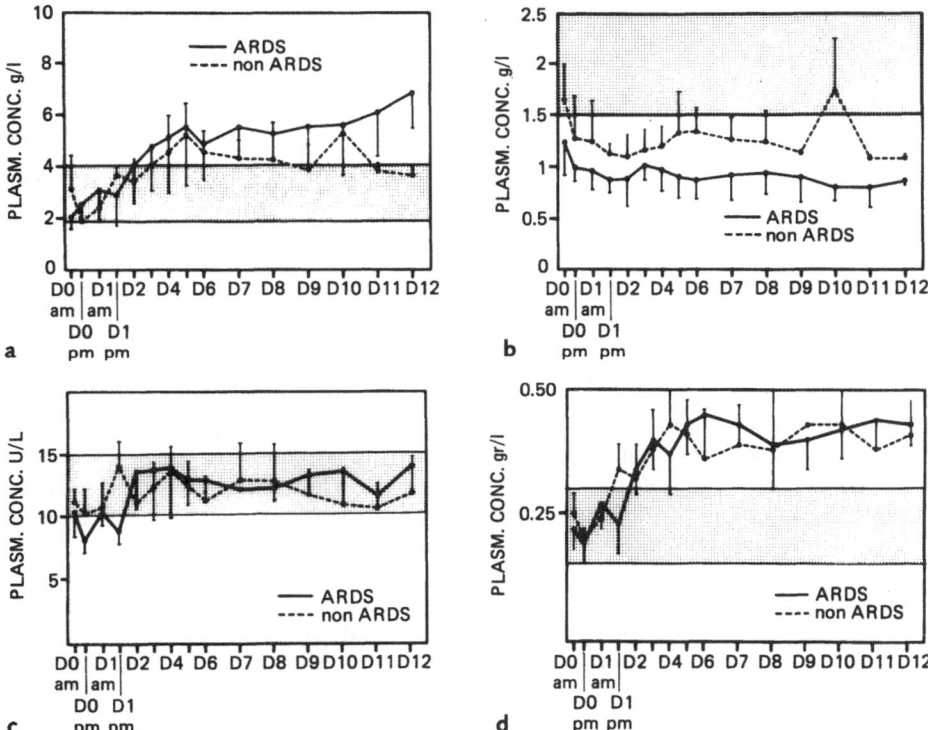

Fig. 5.3a–d. Changes in plasma concentrations of four proteinase inhibitors in 20 critically ill patients over 12 days: **a** α_1-proteinase inhibitor, **b** α_2-macroglobulin, **c** antithrombin III, **d** C1 esterase inhibitor. The *stippled areas* represent the normal range of values in healthy volunteers. D0 = day of injury or surgery.

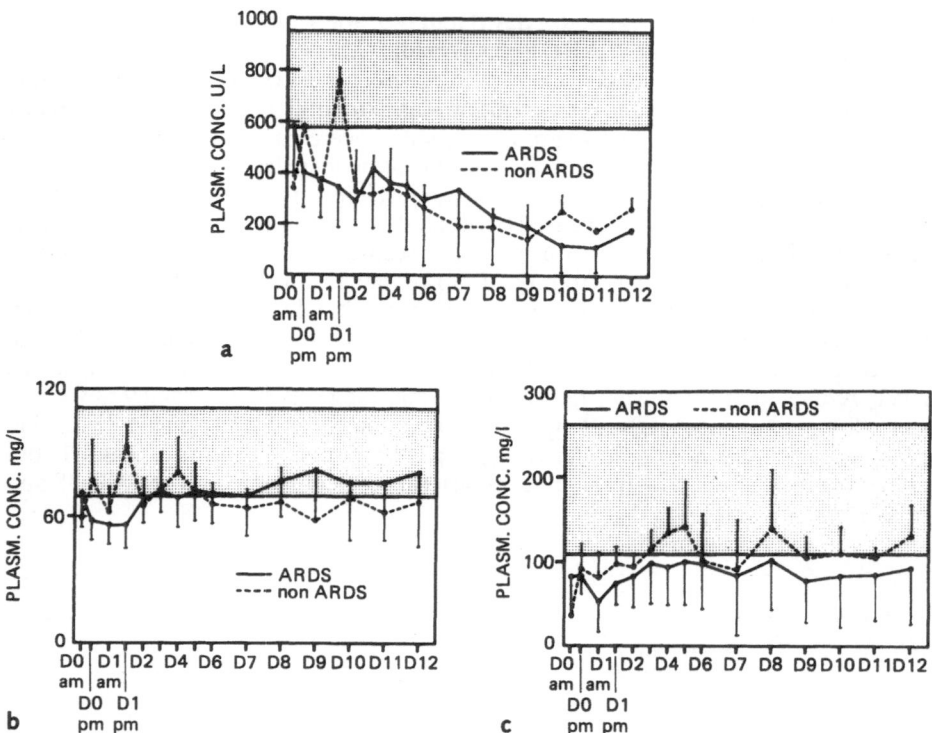

Fig. 5.4a–c. Changes in plasma concentrations of three compounds of the kallikrein–kininogen–kinin system in 20 critically ill patients over 12 days: **a** prokallikrein, **b** high molecular weight kininogens, **c** low molecular weight kininogens. The *stippled areas* represent the normal range of values in healthy volunteers. D0 = day of injury or surgery.

In Fig. 5.4 mean changes in plasma concentration of the different constituents of the KKK system are shown. Prokallikrein activity was constantly below the normal range in both groups; a continuous decrease in plasma concentration was also observed, but there were no significant differences between the two groups. High molecular weight kininogens were lower in ARDS patients and below normal on days 0 and 1, but later returned to low normal levels. Concentrations of low molecular weight kininogens were found to be low in both groups on days 0, 1 and 2. Mean values always remained below the normal range in ARDS patients.

Changes in concentrations of acute phase proteins are shown in Fig. 5.5. In the ARDS group C-reactive protein concentrations were normal on day 0 but increased tremendously, reaching very high levels on days 4 and 5, decreasing slightly until day 9, and then increasing again to even higher concentrations than on day 5. In non-ARDS patients similar changes were observed from day 0 p.m., but peak concentrations were reached at day 1 p.m. with no secondary increase. Concentrations of α_1 acid glycoprotein were in the normal range on days 0 and 1, and increased in the same way in both groups from day 2. After the tenth day they remained high in ARDS patients but decreased slowly in the other group. Haptoglobin concentrations were in the low-normal range from day 0 p.m. to day 4, and then increased to the

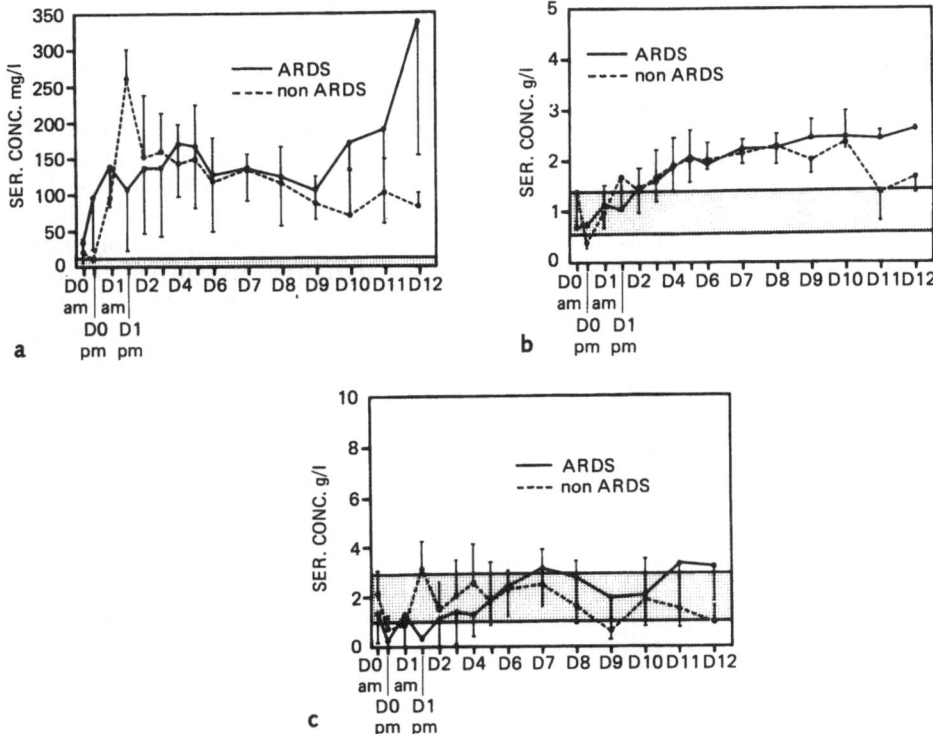

Fig. 5.5a–c. Changes in serum concentrations of three acute phase proteins in 20 critically ill patients over 12 days: **a** C-reactive protein, **b** α_1 acid glycoprotein, **c** haptoglobin. The *stippled areas* represent the normal range of values in healthy volunteers. D0 = day of injury or surgery.

upper limit of normal in ARDS patients. In the non-ARDS patients haptoglobin concentrations remained more or less in the normal range throughout the study.

Discussion

We have demonstrated that all the biochemical parameters measured were abnormal in our critically ill patients: IRT concentrations were increased; antiproteases were decreased (at least at the beginning for α_1PI), the constituents of the KKK system were below normal and there was an increase in acute phase proteins.

The possible interactions between protease–antiprotease and KKK systems are shown in Fig. 5.1. Trypsin, when liberated in the plasma, may induce an activation of the KKK system leading to the production of kinins, which have a detrimental effect on the vascular system (vasodilation and increased permeability of capillaries). Activation of the Hageman factor XII by negatively charged surfaces (collagen, basement membrane, bacterial lipopolysaccharides) occurring in critically ill patients may also induce an activation of kininogens with liberation of kinins, as well as leading to disturbances in the coagulation system. Plasma antiproteases, however, can neutralize proteases (kallikrein, plasmin, trypsin) and constitute a protective mechanism against increased proteolytic activity.

In order to explain the different changes in plasma concentrations of all these measured substances in the critically ill, it is important to be aware of two limitations of our study. Firstly there are the problems surrounding the assay techniques for these substances and, secondly, difficulties in the interpretation of plasma levels in patients in whom there may well be changes in the rate of metabolism. The radioimmunoassay for IRT measures trypsinogen, potential free trypsin and trypsin bound to $\alpha_1 PI$, but it does not measure trypsin bound to $\alpha_2 M$. When released into blood, trypsin is quickly bound by antiproteinases; $\alpha_1 PI$ inhibits the enzymic activity of trypsin while $\alpha_2 M$ binds trypsin without inhibiting its enzymic activity on low molecular weight substrates. Thus, our radioimmunological technique of trypsin measurement does not give us information concerning the presence or absence of trypsin–$\alpha_2 M$ complexes, which may represent enzymically active trypsin.

Our data show that, in critically ill patients, there is a marked and prolonged elevation in IRT, lipase and amylase levels. An important question concerning the origin of these enzymes is whether they are of pancreatic origin. As far as we are aware there is no evidence from the literature that trypsin originates from tissues or cells other than from the pancreas. In a previous report it was shown that IRT was absent in pancreatectomized patients (Felber and Bambule-Dick 1980). The same is true for lipase enzyme liberation. Thus, one may postulate a pancreatic origin for these increases in IRT and lipase. This observation is surprising as there was no clinical suggestion of acute pancreatitis in these patients. We suggest that the pancreas, like other organs, may have suffered from anoxia at the time of injury with shock. This leads to the concept of "shock pancreas", and indeed several studies have shown that the pancreas is highly susceptible to anoxia caused by either decreased arterial PO_2 (Broadie et al. 1979) or reduced blood flow (Warshaw and O'Hara 1978).

For most of our patients who developed ARDS the elevation of pancreatic enzymes followed the development of acute respiratory failure, with the implication that the pancreatic disorder was not responsible for the deterioration of lung function but was accompanied or followed by it. Hammerschmidt et al. (1981) have demonstrated the accumulation of granulocytes in mesenteric vessels in animals after activation of the complement system with zymosan. Activation of the complement system described in ARDS could simultaneously induce lung and pancreas dysfunction (Jacob et al. 1980). An alternative explanation of how the pancreas is damaged in ARDS might be related to the possible defect in the lung's metabolic role. The activity of angiotensin converting enzyme (ACE) is inhibited by acute hypoxia as shown by Leuenberger et al. (1978). ACE (also called kininase II) is known to catabolize kinins which may alter the microcirculation of the pancreas. This hypothesis is unlikely because high plasma concentrations of pancreatic enzymes were found in patients who did not develop ARDS. On the other hand, pancreatic enzymes and kinins could contribute to further deterioration of lung function.

In this study we also found a decrease in plasma concentrations of prokallikrein activity and kininogens of high and low molecular weights. To date there has been no immunological method of bradykinin measurement available, only biological methods. In a recent paper Adam et al. (1986) described the physiological relevance of radioimmunoassay measurements of kininogens. They showed in vitro that a decrease in plasma concentration of kininogens concomitant with a decrease in prokallikrein plasma activity was always associated with kinin liberation. Our data strongly suggest an activation of the kinin system with liberation of kinins. However, there is no firm evidence that kinins are responsible for the development of ARDS. Nevertheless, changes in the plasma concentrations of KKK components appear earlier than changes in IRT, amylase or lipase. This suggests that the kallikrein–kininogen system is activated specifically by the Hageman factor in the early stages, and afterwards by pancreatic and leucocytic enzymes.

Plasma contains several proteinase inhibitors that serve to confine and terminate proteolytic events. In our group of critically ill patients we found very low levels of $\alpha_2 M$ in both ARDS and non-ARDS patients. It must be pointed out that $\alpha_2 M$ is normally present in high concentrations in plasma and is capable of entrapping nearly all known endopeptidases, thereby facilitating their removal from the extracellular space and degradation by mononuclear phagocytes. Our observation that plasma levels of $\alpha_2 M$ were extremely low in critically ill patients suggests either a consumption of this inhibitor or a loss of immunodetectability when bound to proteinases. This consumption remained stable throughout the observation period, but plasma levels were always lower in ARDS patients, perhaps indicating a substantially greater protease release.

AT III plays a major role in controlling serine proteinases in the coagulation cascade. In particular, it inactivates thrombin, but human AT III also inactivates factors IXa, Xa and XIa, plasmin and plasma kallikrein, as well as trypsin and chymotrypsin. For these enzymes, however, the inhibitory effect of AT III has not yet been investigated (Travis and Salvesen 1983). In our critically ill patients we found AT III plasma concentrations at the lower limit of the normal range, except in ARDS patients on day 0 p.m. and day 1 p.m. The coagulation system may have been activated by trauma and sepsis in these patients, explaining the relatively low plasma levels of AT III. However, kinetic studies of inactivation of other enzymes by AT III suggest a rather slow process, and if a similar situation exists for protease–AT III interaction, then it is probable that such inactivation has only a minor role in the control of protease activity.

C1-inh was identified originally as an inhibitor of the first component of complement C1r and C1s. In addition, plasma kallikrein, factor XIa and XIIa and plasmin are inactivated by this inhibitor. Trypsin is not inhibited by C1-inh, but produces a proteolytically modified form of the inhibitor, unable to react with either C1s or plasma kallikrein (Travis and Salvesen 1983). C1-inh shares its role as a primary inhibitor of plasma kallikrein with $\alpha_2 M$; although the rate of inactivation by C1-inh is greater, the concentration of $\alpha_2 M$ in normal plasma is higher, thus offsetting its slower rate of inhibition. C1-inh is presumed to be pivotal in the control of activation of the classical pathway of complement. In a recent study Duchateau et al. (1984) showed significant complement activation in critically ill patients. The low plasma concentration of C1-inh that we observed initially may thus be explained by the consumption of this inhibitor in ARDS as well as in non-ARDS patients, especially early after injury. Later in the course of our study these levels increased and C1-inh behaved as an acute phase protein.

The $\alpha_1 PI$, a broad-spectrum inhibitor which is thought to be especially important in binding neutrophil elastase, may be important in permitting extracellular proteolysis. Recent studies have shown that oxidation inactivates $\alpha_1 PI$. Because leucocytes produce and release several reactive oxygen species during activation, oxidative inactivation of proteinase inhibitors may occur (Travis and Salvesen 1983). In the present study we measured not only free $\alpha_1 PI$, but also $\alpha_1 PI$ complexed with proteases or oxidized α_1PI. We found low plasma levels of α_1PI on days 0 and 1 in the two patient groups, suggesting a consumption of this inhibitor, but afterwards α_1PI increased slowly and behaved as an acute phase reactant.

When looking at the acute phase proteins in these patients we found an early and important increase in C-reactive protein (CRP) over the first 5 days; these levels then decreased but increased again in ARDS patients. With normal convalescence, CRP levels peak at 48 hours and then decline to normal over the ensuing 4 or 6 days. CRP appears to play an important role in the body's defence system after injury. This role is that of an opsonic substance sensitizing damaged cells for cytolysis and removal. This time course may represent maturation of the local tissue injury/repair process, with a decrease in macrophage activation and a fall in interleukin I. But septic complications are heralded by a failure of CRP levels to fall or a recurrent CRP elevation. Sepsis, or a second insult such as surgery, provokes a new round of stimulation of macrophage activity and CRP response as we found in our patients, especially those with ARDS. It also suggests good liver function and metabolism in these critically ill patients.

When looking at changes in the plasma concentration of haptoglobin, we found normal but low levels in the two patient groups. In ARDS patients these levels increased on the fourth day and haptoglobin behaved as an acute phase protein. It is a sensitive acute phase reactant the function of which appears to be to combine with free haemoglobin. The complex is rapidly removed in the liver with consequent conservation of haemoglobin iron that would otherwise be lost through the kidney. This is a method for conserving iron from haemolysis which occurs during trauma, massive transfusion and inflammation. Orosomucoid or α_1 acid glycoprotein is an acute phase reactant which increased in our two groups 48–72 hours after onset of trauma or surgery. This increase persisted in ARDS patients. No clearcut function has been identified for this protein as yet, though the evidence suggests that it may inhibit cathepsin, and may also be involved in platelet function.

In conclusion, our biochemical investigations in critically ill patients at risk of developing ARDS give no evidence for a specific mediator in acute respiratory failure. There is no direct relationship between ARDS and protease–antiprotease imbalance, activation of the kinin system or the inflammatory response. However, there seems to be a more severe disturbance in patients developing ARDS. This may be due either to a greater reactivity of some patients to trauma or even to a pre-existing deficit in their protection system (e.g. antiproteases) or to a combination of the two. There seems to be a close relationship between all these systems but no one of them emerged as a reliable predictor of ARDS.

What about the role of the pancreas? The pancreas probably does not play a role in itself, but may contribute to the occurrence and to the degree of severity of acute respiratory failure. The systemically released pancreatic enzymes are not reliable markers of early evolution of the disease, but they may represent secondary mediators for enhancement of endothelial cell damage and increased permeability.

References

Adam A, Albert A, Calay G, Closset J, Damas J, Franchimont P (1985a) Human kininogens of low and high molecular mass: quantification by radioimmunoassay and determination of reference values. Clin Chem 31:423–426

Adam A, Ers P, Boulanger J, Albert A, Azzouzi M, Faymonville ME (1985b) Optimized determination of plasma prokallikrein on a Hitachi 705 Autonanalyser. J Clin Chem Biochem 23:203–207

Adam A, Damas J, Ers P, Albert A, Stas JL, Lecomte J (1986) Pathophysiological meaning of quantification of human kininogens by radioimmunoassay. Pathol. Biol. 34:19–24

Adam A, Albert A, Boulanger J, Genot D, Demoulin A, Damas J (1985) Influence of oral contraceptives and pregnancy upon plasma constituents of the kallikrein–kininogen system. Clin Chem 31:1533–1536

Broadie TA, Devedas M, Rysavy J et al. (1979) The effect of hypoxia and hypercapnia in canine pancreatic blood flow. J Surg Res 27:114–118

Deby-Dupont G, Haas M, Pincemail J, Braun M, Lamy M, Deby C, Franchimont P (1984) Immunoreactive trypsin in the adult respiratory distress syndrome. Intensive Care Med 10:7–12

Duchateau J, Haas M, Schreyen H, Radoux L, Sprangers I, Noel FX, Braun M, Lamy M (1984) Complement activation in patients at risk of developing the adult respiratory distress syndrome. Am Rev Respir Dis 130:1058–1064

Felber JP, Bambule-Dick J (1980) Radioimmunoassay of plasma trypsin in pancreatic diseases and in juvenile-onset diabetes. In: Podolsky S, Viswanathan M (eds) Secondary diabetes: the spectrum of the diabetic syndromes. Raven Press, New York, pp 191–196

Geokas MC, Largman C, Brodrick JW, Johnsen JH (1979) Determination of human pancreatic cationic trypsinogen in serum by radioimmunoassay. Am J Physiol 236:E77

Hammerschmidt DE, Harris PD, Wayland JH et al. (1981) Complement-induced granulocyte aggregation in vivo. Am J Pathol 102:146–150

Jacobs HS, Craddock PR, Hammerschmidt DE, Moldow CF (1980) Complement-induced granulocyte aggregation. an unsuspected mechanism of disease. N Engl J Med 302:789–794

Leuenberger PJ, Stalcup SA, Mellins RB et al. (1978) Decrease in angiotensin 1 conversion by acute hypoxia in dogs. Proc Soc Exp Biol Med 158:586–589

Malvano R, Marchisio M, Massaglia A (1980) Radioimmunoassay of trypsinlike substance in human serum. Scand J Gastroenterol 15 [Suppl 62]:3

Nicod PH, Leuenberger C, Seydoux F, Rey F, Van Melle G, Perret C (1985) Evidence for pancreas injury in adult respiratory distress syndrome. Am Rev Respir Dis 131:696–699

Pontoppidan H, Huttemeier PC, Quinn BA (1985) Etiology, demography and outcome. In: Zapol W (ed) Acute respiratory failure. Falke, New York, pp 1–21

Travis J, Salvesen GS (1983) Human plasma proteinase inhibitor. Ann Rev Biochem 53:655–709

Warshaw AL, O'Hara PJ (1978) Susceptibility of the pancreas to ischemic injury in shock. Ann Surg 188:197–201

Chapter 6

Changing Haemodynamic Concepts in Human Septic Shock

L. G. Thijs, A. B. J. Groeneveld and A. J. Schneider

Septic shock is associated with complex alterations in haemodynamic variables and cardiovascular performance. Over the last 20 years much information derived from clinical studies has been accumulated. However, uncertainty still exists as to the exact pathophysiological mechanisms underlying these haemodynamic changes which make septic shock such a highly lethal syndrome (more than 50% mortality). Of all patients with sepsis approximately 40% develop profound hypotension and shock. Although it has been appreciated in the last decade that the severity of the underlying disease is a major factor determining ultimate outcome (Kreger et al. 1980; McGabe 1974; Thijs et al. 1984), a thorough understanding of the various cardiovascular changes in septic shock remains essential for appropriate treatment. Together with aggressive therapy of the infectious process with high-dose antimicrobials and surgical eradication of a septic source when necessary, improvement of tissue oxygenation is a prerequisite for survival. Antibiotic treatment as such cannot be expected to affect mechanisms that have already been activated and which deeply disturb systemic haemodynamics, the control mechanisms of the microcirculation, tissue oxygenation and, as a result, cell function. Methods for maintaining or improving tissue oxygenization include optimal pulmonary and circulatory management in order to optimize oxygen delivery. The way this can best be achieved largely depends on the concepts available concerning the nature of the cardiovascular derangements.

One of the earliest clinical studies (Udhoji and Weil 1965) suggested that cardiac output in patients with bacteraemic shock was reduced. This finding is in disagreement with almost all subsequent clinical studies, which have uniformly demonstrated that patients with septic shock as a rule have a normal or high forward flow and a lowered peripheral vascular resistance (Gunnar et al. 1973; Hess et al. 1981; Siegel et al. 1967; Winslow et al. 1973; Wilson et al. 1965, 1967, 1971). The discrepancy between the early and later studies can probably best be explained by differences in cardiovascular volume status, i.e. in the early studies patients had both sepsis and uncorrected hypovolaemia (Wilson et al. 1967, 1971).

Prospective studies during genito-urinary procedures suggest that the first haemodynamic alteration in sepsis is a decrease in systemic vascular resistance (Blain et al. 1970; Gunnar et al. 1973; Robinson et al. 1975). This is usually accompanied by a reflex increase in cardiac output to maintain arterial pressure. When the patient is unable to increase forward flow appropriately, blood pressure falls and ultimately a shock syndrome ensues. This may be the result of either coexisting (relative) hypovolaemia, impaired myocardial function or an extremely pronounced peripheral vasodilation. Septic shock is therefore, at the onset, generally charac- terized by a hyperdynamic circulation (when hypovolaemia is corrected) with an elevated cardiac index, hypotension and decreased systemic vascular resistance (Gunnar et al. 1973; Hess et al. 1981; Parker et al. 1984; Siegel et al. 1967; Winslow et al. 1973; Wiles et al. 1980; Wilson et al. 1965, 1967, 1971).

In the early literature haemodynamic differences between gram-positive and gram-negative septic shock appeared to be an important controversy (Blain et al. 1970; Gunnar et al. 1973; Kwaan and Weil 1969; Winslow et al. 1973). It is now widely recognized, though, that while there may be some differences between these two groups, the haemodynamic response to bacteraemia is essentially independent of the nature of the causative organism (Parker et al. 1984; Weisel et al. 1977; Wiles et al. 1980; Wilson et al. 1971).

The inability to increase blood flow efficiently has been associated with a poor out- come. In one study (Abraham et al. 1984) a difference in cardiac index (CI) between survivors and non-survivors was already observed in the 24-hour period prior to the shock episode. Several investigators have shown a relationship between a low cardiac output at the onset of management and a high mortality (Bell and Thal 1970; Hess et al. 1981; Nishijma et al. 1973; Weil and Nishijma 1978). However, others including our own group (Groeneveld et al. 1986) have failed to show a difference in initial CI between survivors and non-survivors (Weisel et al. 1977; Winslow et al. 1973). This discrepancy may be the result of a number of factors. Initial cardiac output is influ- enced by coexisting hypovolaemia which was often not adequately ruled out in earlier studies as an appropriate measure of left ventricular filling was not available at the time. In the recent past, only central venous pressure (CVP) was measured, which is an inadequate substitute for pulmonary capillary wedge pressure (PCWP) for assess- ing left ventricular filling as CVP and PCWP do not necessarily correlate in septic shock (Calvin et al. 1981a; Krausz et al. 1977). Furthermore, the presence or absence of myocardial disease prior to the septic insult has not always been adequately assess- ed. On the other hand, chronic liver disease is usually characterized by a high cardiac output and supervening septic shock in these patients carries an especially poor prognosis (Baumgartner et al. 1984; Bell and Thal 1970). Moreover, patients may have been studied at different stages of their disease, which may explain the heterogeneity of the results in the patients investigated. Nevertheless, there seems to be some sense in the theory that patients with compromised cardiac function are less able to cope with the burden imposed on the circulation by the septic insult and there- fore have a less favourable outcome. However, no comprehensive study supporting this theory is yet available (Parillo 1985).

Although there are only a few studies in which serial haemodynamic measure- ments have been performed during the course of septic shock, it is widely assumed that in non-survivors the hyperdynamic state slowly evolves into a hypodynamic state (Clowes et al. 1970, 1975; Hess et al. 1981; Kwaan and Weil 1969; Nishijma et al. 1973; Shoemaker 1971; Siegel et al. 1967; Weil and Nishijma 1978; Weisel et al. 1977). In other words some patients develop a decrease in cardiac output during their later clinical course which ultimately leads to their death. However, a limited number of reports do describe patients dying in a hypodynamic state, i.e. CI <2.5 l min^{-1} m^{-2}, hypotension, increased peripheral resistance and progressive lactacidaemia (Clowes et al. 1970; Kwaan and Weil 1969; Nishijma et al. 1973; Udhoji and Weil 1965). Although only CVP was used to measure the adequacy of cardiovascular volume status, the findings have been interpreted as a sign of myocardial failure.

These observations have initiated a widespread and continuing discussion about myocardial depression and cardiac failure in septic shock (Ellrodt et al. 1985; Gunnar et al. 1973; Hess et al. 1981; Parillo 1985; Parker et al. 1983; Siegel et al. 1967; Vincent et al. 1981; Weisel et al. 1977; Winslow et al. 1973). The issue of intrinsic myocardial dysfunction has been the subject of numerous experimental studies (Goldfarb 1982; Parker and Adams 1985). The major question is whether cardiodynamics during septic shock reflect an underlying dysfunction of intrinsic myocardial contractility, or simply the adjustment of an otherwise normal heart to the abnormal environment in which it is forced to contract (Parker and Adams 1985). In situ studies of myocardial behaviour in endotoxic or septic shock models have failed to reach a consensus (Goldfarb 1982; Parker and Adams 1985). This may be explained in part by the difficulty of defining and assessing contractility in an in situ preparation—as indicated by the large variety of indices used to describe myocardial function (Goldfarb 1982; Parker and Adams 1985).

The lack of consensus between experimental investigators casts doubt upon the feasibility of assessing myocardial contractility without the influence of pre- and afterload in the clinical situation. Nevertheless, an impressive number of clinical observations suggest that depression of myocardial function does occur in human septic shock. The cardiac response to volume loading expressed as increase in CI and left ventricular stroke work index was smaller in non-survivors than in survivors in some studies (Vincent et al. 1981; Weisel et al. 1977).

Weisel et al. constructed left ventricular performance curves (Frank–Starling curves) during fluid administration in patients with sepsis and demonstrated that non-survivors showed significantly lower left ventricular stroke work (both maximum and slope of increase) at comparable left ventricular filling pressures than did survivors. Volume loading in 40 patients with septic shock resulted in a more than 20% increase in cardiac output in only 18 patients, while in 8 patients cardiac output fell despite an increase in left ventricular filling pressure (Winslow et al. 1973). This response could not predict survival but was considered as evidence of impaired myocardial function.

Measurement of Cardiac Volume and Ejection Fraction

A new dimension was added to the study of myocardial behaviour in septic shock with the introduction of bedside gated blood pool scanning. This technique offers the possibility of measuring ejection fraction and of approximating end-systolic and end diastolic ventricular volumes. It was soon appreciated that in septic shock CVP and PCWP did not correlate with right and left ventricular end diastolic volumes (RVEDV and LVEDV) respectively (Calvin et al. 1981b; Ellrodt et al. 1985; Hoffman et al. 1983; Kimchi et al. 1984). In a study of 13 patients in early septic shock we also failed to find such a correlation (Fig. 6.1). This points to a change in ventricu-

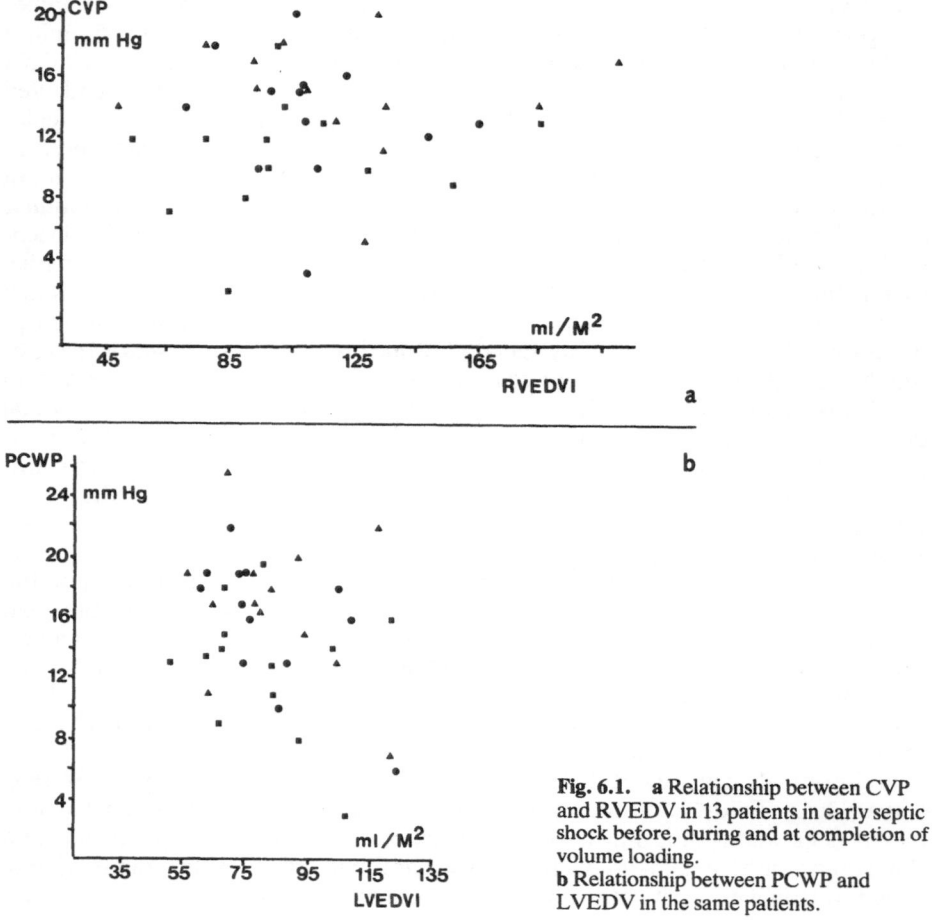

Fig. 6.1. **a** Relationship between CVP and RVEDV in 13 patients in early septic shock before, during and at completion of volume loading.
b Relationship between PCWP and LVEDV in the same patients.

lar diastolic compliance in this condition. Ventricular end-diastolic volume is a better reflection of ventricular preload than filling pressure (Calvin et al. 1981b). The availability of this technique may therefore contribute to a better understanding of myocardial performance in septic shock.

In a series of 20 patients with sepsis but without evidence of shock and who were early in the course of their disease Calvin et al. (1981b) showed that left ventricular ejection fraction (LVEF) was comparable with that in a control group (0.54 ± 0.03 versus 0.50 ± 0.02). Furthermore, average dv/dt, representing velocity of ejection, was higher in sepsis compared with the controls. The authors concluded that depression of myocardial contractility is not a major feature of early human sepsis. However, in 5 out of their 20 patients LVEF was less than 0.40, which was associated with a larger mean LVEDV than in the remaining 15 patients (142.3 ± 4.7 versus 86.2 ± 4.9 ml m^{-2}) while the mean PCWP was not higher in these patients (10.3 ± 1.8 versus 11.1 ± 2.1 mmHg).

In 1984 Parker et al. reported the results of a serial study in 20 patients with documented septic shock that had used radionuclide cineangiography. All patients had a normal or elevated CI. The group as a whole initially had a low mean LVEF of 0.40 ± 0.04 which gradually rose over 7–10 days to 0.55 ± 0.05. Survivors were found to have a lower initial LVEF (0.32 ± 0.04) than non-survivors (0.55 ± 0.03). In survivors LVEF remained low for up to 4 days and then returned to normal (0.55 ± 0.05), while in non-survivors LVEF remained above or in the normal range to within 24 hours before death. In three surviving patients radionuclide angiographic studies had also been performed prior to the onset of septic shock. All three patients showed a fall in ejection fraction from their baseline values with full recovery after 7–10 days. In this study it was also found that survivors initially had an acutely dilated left ventricle with a large end diastolic volume index (159 ± 29 ml m^{-2}) which decreased to normal values (72 ± 12 ml m^{-2}) in 7–10 days. In contrast, the non-survivors had a normal initial LVEDV (81 ± 9 ml m^{-2}) which did not change with time.

A possible explanation for these findings may be that all patients with septic shock develop myocardial depression. Since the non-survivors have a lower systemic vascular resistance than survivors, this lower afterload may result in a normal LVEF (and LVEDV) despite reduced myocardial contractility. An alternative explanation for the absence of LV dilation in the non-survivors might be the development of myocardial oedema with a resulting loss of myocardial compliance and inability of the left ventricle to dilate. The survivors may have had less myocardial oedema, permitting ventricular dilation in response to an unknown stimulus causing myocardial depression, but a higher peripheral resistance, thus maintaining cardiac output (Parker et al. 1984).

Ellrodt et al. (1985) studied 35 patients with culture-proven septic shock, 15 of whom had previously existing heart disease. On presentation 34 patients had at least one abnormality in left ventricular performance. Left ventricular stroke work was depressed in 33 patients. The mean LVEF was 0.44 ± 0.15 but only 16 patients (46%) had a LVEF greater than 0.48. In this study regional wall motion of the left ventricle was also assessed, using nuclear angiography; generalized and segmental abnormalities were found in 4 and 22 patients respectively (74%). These patients had significantly lower LVEF than those with normal wall motion. Surprisingly, the mean LVEF for patients with a history of heart disease was not significantly different from the mean LVEF of those without such a history. The mean LVEDV for patients with global or segmental wall motion abnormalities was greater than in patients with normal values (90 versus 60 ml m^{-2}). In 22 patients follow-up studies were performed after an average of 2.3 days. One patient developed a decrease in LVEF, 5 with a low LVEF had normal values in the second study and 3 patients resolved their segmental wall motion abnormalities. It appears to be important that conventional haemodynamic variables were of no value in predicting those patients with either a depressed LVEF or segmental wall motion abnormalities. The authors concluded that reversible left ventricular dysfunction is common in septic shock and is most often segmental in nature, and that these parameters had no effect on outcome.

In a number of investigations the effect of volume loading on ventricular function was assessed while using ventricular end-diastolic volumes as an index of ventricular preload. A study of 28 critically ill patients without evidence of shock, but a large percentage of whom had sepsis, showed an inhomogenous response to volume loading not primarily related to the underlying illness (Calvin et al. 1981a). In the group as a whole CVP, PCWP and CI increased. However, neither LVEDV nor LVEF showed a significant change in response to fluid challenge. When subdividing the group it was found that 20 patients increased their stroke volume (responders) while 8 patients did not (non-responders). These subgroups showed no difference in baseline values. In the non-responders the mean LVEDV did not change although the mean PCWP increased. It was suggested by the authors that a change in left ventricular compliance and function in this subgroup could be caused by a change in right ventricular loading conditions. The responders increased their stroke volume secondary either to an increase in LVEDV ($n = 11$) or to an increase in LVEF ($n = 9$) secondary to a reduced impedance to ejection of the left ventricle as a result of a decreased systemic vascular resistance. It was found that patients characterized by an increase in LVEDV in response to fluid administration showed a concomitant fall in pulmonary vascular resistance. Patients characterized by an increase in LVEF did not exhibit a change in this parameter. Although this study provides interesting and important observations it is difficult to use these to evaluate myocardial function in sepsis because of the heterogeneity of the patients studied.

However, in another study (Parillo 1985) patients with sepsis or early septic shock demonstrated similar, although not identical, abnormalities in response to volume loading. Some patients developed an increase in left ventricular size, i.e. ventricular dilatation together with an elevated PCWP but without increase in stroke work. This group was considered to have an abnormality in ventricular contractility. Another subgroup failed to develop ventricular dilatation despite volume-induced elevation in PCWP. It was concluded that these patients had a compliance abnormality.

Right Ventricular Failure

In the literature dealing with myocardial function in septic shock the left side of the heart has received the largest amount of attention. However, already in 1970 Clowes et al. suggested that right ventricular failure secondary to pulmonary lesions might contribute to circulatory insufficiency and death. They reported on 38 patients with severe sepsis due to generalized fulminating peritonitis. In those patients in whom right ventricular or pulmonary artery pressure was measured it was found that in the acute phase pulmonary hypertension was present. In the patients who died CI was low while pulmonary artery pressure and resistance remained elevated. The same group (Clowes et al. 1975) showed in a subsequent study that in a number of septic shock patients progression is characterized by an increase in pulmonary vascular resistance and a decrease in CI, again indicating that right ventricular failure might have occurred.

In another study 22 out of 37 patients with sepsis and adult respiratory distress syndrome (Sibbald et al. 1978) showed a significant degree of resistance to flow in the pulmonary circulation. Twelve of these patients had a pulmonary artery diastolic to capillary wedge gradient of greater than 5 mmHg; 10 (83%) died within 2 days. Of 14 patients who were initially without an elevated mean pulmonary artery pressure, 8 subsequently developed pulmonary hypertension. These findings confirm that severe and prolonged bacterial infection is frequently accompanied by pulmonary hypertension.

Assessment of global right ventricular performance in shock may be improved when right ventricular ejection fraction (RVEF) and right ventricular volumes are measured. Hoffman et al. (1983) studied 16 patients after resuscitation from septic ($n = 9$) or hypovolaemic ($n = 7$) shock, using gated cardiac scintigraphy. All patients showed, when studied within 24 hours after resuscitation, a depressed RVEF when compared with controls (septic, 0.35 ± 0.16; hypovolaemic, 0.31 ± 0.05; controls, 0.52 ± 0.07). At the same time RVEDV was significantly elevated (septic, 143 ± 28; hypovolaemic, 106 ± 22; controls, 55.8 ± 7.4 ml m^{-2}). No significant differences were found in LVEF between controls and shock patients, but LVEDV in patients with sepsis was significantly greater when compared with controls (107 ± 39 versus 52.3 ± 17.4 ml m^{-2}). This was not the case in hypovolaemic patients (58.5 ± 16.4 ml m^{-2}). CI was significantly higher in patients with sepsis while systemic vascular resistance was lower. In 10 patients follow-up studies were performed. Improvement in the clinical condition ($n = 7$) was accompanied by an increase in RVEF and a decrease in RVEDV while deterioration ($n = 3$) was characterized by an increase in RVEDV without a significant change in RVEF.

Another recent study using a similar technique concerned 25 patients in septic shock (Kimchi et al. 1984). A depressed RVEF (<0.38) was found in 13 patients (52%) of whom 8 also had a depressed LVEF (<0.48). In the initial study statistical analysis failed to reveal a significant relationship between RVEF and mean pulmonary artery pressure and resistance. In addition, no significant difference was found between mean RVEF in patients with acute respiratory failure and RVEF in patients without this complication. In 17 patients the study was repeated after 1–8 days. Of 13 patients with initial RVEF impairment 3 had normal pulmonary artery pressure and 10 had pulmonary hypertension. All 3 patients without pulmonary hypertension had normal RVEF on repeat study. In contrast, only 3 out of the 10 with pulmonary hypertension showed improvement in RVEF. The authors concluded that right ventricular dysfunction is not an uncommon finding but its presence cannot be predicted on the basis of clinical and conventional haemodynamic data alone.

To study right and left ventricular function in septic shock an experimental model was developed by our group using live *E. coli* sepsis in pigs (Teule el al. 1984, 1985). This model is characterized by early pulmonary hypertension and systemic hypotension. It was found that while cardiac output was maintained, RVEDV increased and LVEDV decreased. A significant increase in RV stroke work was observed, along with a decrease in LV stroke work. Although RVEF tended to decrease, this decrease was not found to be significant. The effect of volume loading was also studied in this model (Schneider et al. 1986). It was reasoned that volume loading in the presence of increased right ventricular afterload might increase RVEDV to such an extent that RV failure would ensue together with reduction in LVEDV and impairment of left ventricular function as a consequence (Calvin et al. 1981a). However, we detected a similar response of both ventricles: RVEDV and LVEDV both increased and their ratio did not change. When ventricular performance curves were plotted it was found that both right and left ventricular function were depressed.

As volume loading is an important initial step in the treatment of human septic shock the similar response of both ventricles is an important observation. It has been suggested that aggressive volume therapy in the presence of pulmonary hypertension may be responsible for right ventricular dilatation, a left shift of the intraventricular septum with a decrease in left ventricular volume and compliance, and fall in cardiac output (Calvin et al. 1981a). If this is true it may have consequences for the management of septic shock patients. We therefore studied the effect of volume loading (500 ml plasma in 30 minutes) in early septic shock using gated blood pool scanning. A

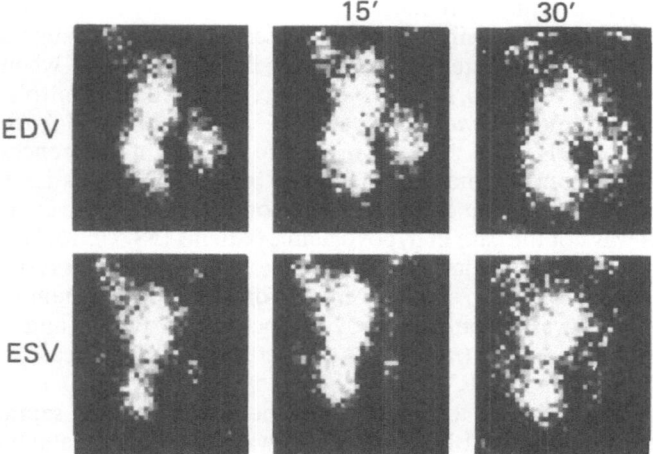

Fig. 6.2. Gated blood pool scan showing the effect of volume loading on right and left ventricular end-diastolic (EDV) and end-systolic (ESV) volume. In this patient CI did not increase, RVEDV increased, and LVEDV did not change or decreased slightly.

total of 13 patients was investigated. Before volume loading mean pulmonary arterial pressure was 23.6 ± 4.7 (SD) mmHg, mean RVEDV 103 ± 36 ml m^{-2}, mean RVEF 0.50 ± 0.17, mean LVEDV 81 ± 20 ml m^{-2} and mean LVEF 0.60 ± 0.09. None of the patients had a LVEF smaller than 0.50; in 6 it was between 0.50 and 0.55. RVEF was smaller than 0.40 in 4 patients. During volume loading mean CI increased from 5.3 ± 1.5 to 5.8 ± 1.6 l min^{-1} m^{-2} while blood pressure normalized and systemic and pulmonary vascular resistance decreased although pulmonary arterial pressure increased slightly. RVEDV increased to 121 ± 45, LVEDV to 84 ± 20 ml m^{-2}. Mean RVEF decreased slightly to 0.48 ± 0.16, LVEF increased slightly to 0.63 ± 0.10.

On analysis of individual cases it was found that LVEDV decreased in 6 patients but in only 1 of these had RVEDV increased (Fig. 6.2). Only 8 patients increased their CI significantly, 5 by an increase in LVEDV, 2 by an increase in LVEF and 1 by both mechanisms (Figs. 6.3, 6.4). In the 5 non-responders RVEDV increased in 2 while CVP increased in all, and LVEDV decreased slightly in 4 while PCWP rose in all except one (Figs. 6.2, 6.3). The main preliminary conclusion from this study is that failure to respond to volume loading in early septic shock is, as a rule, not caused by compression of the left ventricle due to overdistension of the right ventricle.

Pathogenesis of Myocardial Depression in Septic Shock

This review of the literature provides sufficient evidence for myocardial depression being a mechanism in septic shock. It also shows the limitations of conventional haemodynamic measurements for assessing myocardial performance. There has been much speculation as to the exact nature of the pathogenetic mechanisms that lead to depressed myocardial function. Coronary hypoperfusion and myocardial

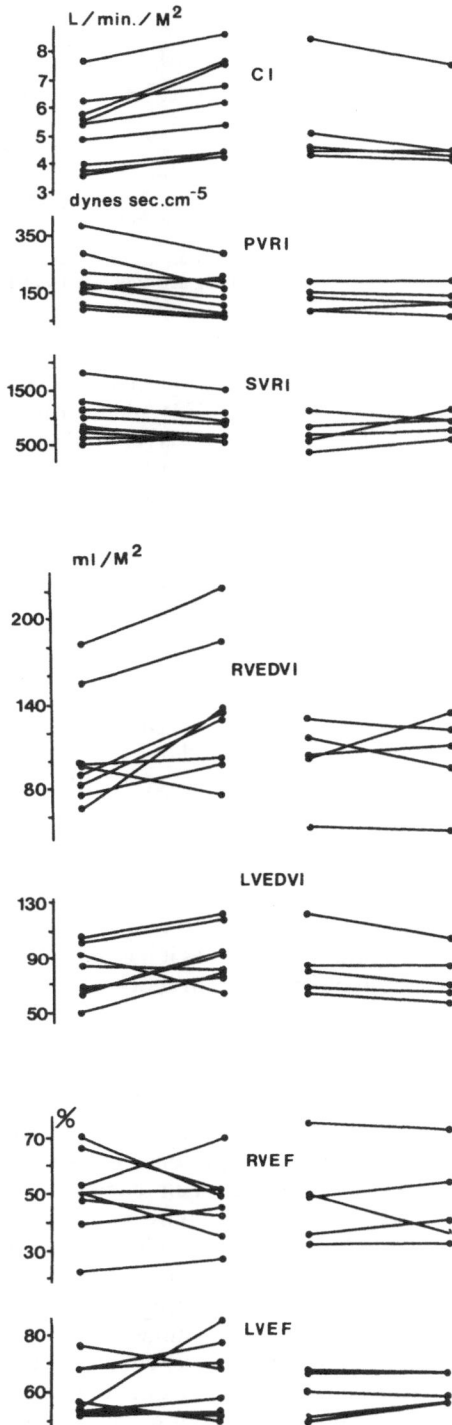

Fig. 6.3. Effect of volume loading on CI, pulmonary and systemic vascular resistance (PVR, SVR), RVEDV, LVEDV, RVEF and LVEF showing individual data. Left-hand panel shows data of patients who increased their CI, right-hand panel the data of patients without such an increase.

15'   30'

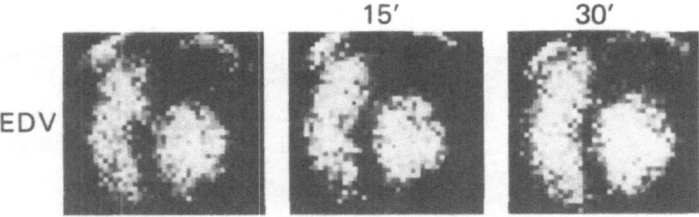

EDV

Fig. 6.4. Gated blood pool scan showing increase in both right and left ventricular end-diastolic volume (EDV) during volume loading accompanied by an increase in CI.

ischaemia have been suggested as major factors (Hess et al. 1981; Hinshaw et al. 1974; Hoffman et al. 1983), mainly on the basis of data from animal studies. In one study of patients with sepsis (Ellrodt et al. 1985), however, no relationship was found between coronary artery perfusion pressure and left ventricular stroke work, ejection fraction or frequency of left ventricular segmental wall dysfunction, either in the entire group or the subgroup of patients with coronary artery disease. This suggests that if myocardial ischaemia is present it does not represent a simple reduction in coronary perfusion pressure.

In another study, using a thermodilution coronary catheter (Parillo 1985), it was found that in patients with septic shock who had decreased LVEF, coronary blood flow was not reduced and lactate extraction was unchanged. Nevertheless, marked redistribution of flow within the myocardium may occur with a resulting ischaemia of the subendocardial layers (Bronsveld et al. 1985; Weisel et al. 1977). Local coronary vasoconstriction induced by humoral factors such as serotonin, noradrenaline, thromboxane A_2 (Reines et al. 1982) and angiotensin (Hess et al. 1981) may be responsible for myocardial segmental abnormalities (Ellrodt et al. 1985). Arterial hypoxaemia and reduced oxygen unloading of the red cells may be factors contributing to myocardial ischaemia. In a series of patients with sepsis Weisel et al. (1977) demonstrated that a reduced myocardial performance was more pronounced in patients with a left-shifted oxygen dissociation curve.

In experimental studies it has repeatedly been shown that a myocardial depressant factor can be produced in shock, possibly from the ischaemic splanchnic area (Lefer 1979; Maksad et al. 1979; Nespoli et al. 1983). Recently, it has been demonstrated that blood obtained during the acute phase of septic shock contains a substance capable of depressing myocardial contraction. The decrease in contraction amplitude in vitro correlated with the depressed LVEF found in vivo. It was not found in other critically ill patients nor in patients who had recovered from septic shock (Parillo 1985). Although the presence of myocardial depressant substances has been sufficiently substantiated, their clinical significance has not yet been demonstrated beyond any doubt. Not only myocardial systolic contractile behaviour seems to be impaired, but changes in myocardial diastolic behaviour have also been observed. A number of factors can induce a decrease in myocardial diastolic compliance, such as ischaemia and inotropic drugs. A decrease in the left ventricular volume–pressure relationship may also be induced by right ventricular dilatation (Calvin et al. 1981a; Hoffman et al. 1983; Kimchi et al. 1984). An additional explanation for loss of compliance might be the occurrence of myocardial oedema (Parker et al. 1984; Weisel et al. 1977). In a study by our group (Bronsveld et al. 1985) on canine endotoxic shock the development of intracellular myocardial oedema with compression of the capillaries was observed, which has been confirmed by other experimental studies (Postel and Schloerb 1977). Oedema may affect not only compliance but also systolic contractile behaviour. Several studies suggest that a variety of metabolic factors, including low insulin levels, are involved in the pathogenesis of myocardial dysfunction in septic shock (Clowes et al. 1978; Siegel et al. 1979).

In some patients increased left ventricular afterload may be a pathogenetic factor (Cerra et al. 1978), but other studies have failed to correlate systemic vascular resistance and LVEF (Ellrodt et al. 1985). Nevertheless, an elevated afterload induced by factors such as catecholamines and angiotensin (Hess et al. 1981) may contribute to depressed left ventricular performance in some patients.

An increase in pulmonary vascular resistance poses an increased pressure load on the right ventricle, which dilates in order to maintain stroke volume. This compensatory mechanism results in increased right ventricular wall stress and increased oxygen requirements. This makes the right ventricle vulnerable to diminished coronary blood flow (Hoffman et al. 1983). Resulting ischaemia may then induce right ventricular failure. In addition, other mechanisms already mentioned may affect the right ventricle as well. Blockage of the pulmonary circulation with microthrombi composed of platelets and leucocytes has been demonstrated (Sibbald et al. 1978). Hypoxic pulmonary vasoconstriction also increases pulmonary vascular resistance (Sibbald et al. 1978). Direct effects of endotoxin and neurogenic factors acting through the autonomous nervous system may also be operative, and a number of humoral factors such as histamine, serotonin, catecholamines, angiotensin, bradykinin and prostaglandin F_2 may be involved (Clowes et al. 1970; Hess et al. 1981; Kimchi et al. 1984; Sibbald et al. 1980).

Peripheral Vascular Failure

So far, the occurrence of myocardial depression in septic shock has been discussed. It has been shown that it is an entirely reversible phenomenon in patients who survive (Ellrodt et al. 1985; Parillo 1985; Parker et al. 1984; Weisel et al. 1977). A decreasing cardiac output during the course of the disease is considered an ominous prognostic sign. However, a number of observations in the recent literature indicate that cardiac output does not always fall in cases with fatal outcome (Baumgartner et al. 1984; Groeneveld et al. 1986; Parker et al. 1983; Thijs et al. 1984). In these patients blood pressure falls without a decrease in forward flow, indicating that persistent peripheral vasodilation may be an important mechanism associated with an unfavourable prognosis. It has been proposed that the documented decrease in peripheral vascular resistance at the onset of sepsis and septic shock should be attributed to the opening of arteriovenous shunts (Cohn et al. 1968; Siegel et al. 1979).

In the septic hindlimb model in dogs, a study using labelled microspheres did indeed show increased "anatomical" arteriovenous shunting in the septic area (Cronenwett and Lindenauer 1979). Although some "anatomical" shunting in local septic areas may be present the available evidence suggests that systemic "anatomical" shunting is not significantly changed in sepsis (Finley et al. 1975; Houtchens and Westenskow 1984). In a canine endotoxin model systemic arteriovenous shunts, as measured with microspheres, even decreased (Lambalgen et al. 1984). It is now widely accepted than an abnormally low vascular tone causes the initial fall in systemic vascular resistance. Several mediators such as bradykinin, histamine, vasodilator prostaglandins, endorphins, and products of complement activators as mentioned in Chapter 3, seem to be responsible for this fall (Hess et al. 1981; O'Donnell et al. 1976; Robinson et al. 1975).

Patients with hyperdynamic septic shock show a high cardiac output and a low arterial mixed venous oxygen difference together with elevated blood lactate levels. Although blood flow to the periphery increases the tissues are apparently unable to extract sufficient oxygen and local ischaemia supervenes, resulting in lactate production. Because of the generalized vasodilation the finely tuned adaptation between perfusion and oxygen demand in the tissues seems to be disturbed. This results in maldistribution of flow causing areas with a relatively high perfusion and a relatively low need for oxygen ("physiological" shunting) and a relatively low perfusion in areas that need relatively high amounts of oxygen. Because of this loss of autoregulation a fall in blood pressure may cause ischaemia in the latter areas, a process which is aggravated by aggregation of platelets and leucocytes (Cain 1984).

In a series of 50 patients with culture-proven septic shock two-thirds of the non-survivors had a very low systemic vascular resistance and a normal or high CI within a few hours of death. Only one-third of the non-survivors developed a low CI prior to death (Parker et al. 1983). Already in 1967 Siegel et al. suggested that the poorest prognosis appeared in those patients with the highest CI and the most pronounced peripheral vasodilation. In a recent study conducted in our unit (Groeneveld et al. 1986), in 42 patients with septic shock it was found that initial haemodynamic variables did not differ between survivors and non-survivors. However, the 21 non-survivors showed a progressive decline in blood pressure and a rise in arterial blood lactate levels, while in most cases CI was maintained. The majority of the non-surviving patients remained hyperdynamic until shortly before death. The common denominator of the patients who died of septic shock seems to be persistent vasodilation. These findings suggest that patients with septic shock are more likely to die of peripheral vascular failure than of cardiac failure.

The exact pathogenetic mechanism of this persisting vasodilation and failure to extract oxygen is unknown. Undoubtedly local metabolic and humoral factors are important. Tissue hypoxia and acidosis may induce vasodilation. Adrenergic receptors may be "down-regulated" in the course of the disease which may explain the diminished responsiveness to catecholamines (Groeneveld et al. 1986; Thijs et al. 1984). The similarity in haemodynamic features between patients with liver cirrhosis and those with septic shock prompted some authors to speculate that the same pathophysiological factors may be involved. Impaired hepatic metabolism of aromatic amino acids in both conditions might lead to the production of false neurotransmitters, which may be responsible for a decrease in vascular tone (Nespoli et al. 1983; Siegel et al. 1982). Knowledge of underlying mechanisms is necessary before therapeutic ways of controlling these abnormalities of the peripheral vasculature can be found.

Conclusion

It appears that our concept of the haemodynamic changes and their clinical significance in septic shock is presently undergoing changes. The largest amount of clinical investigational effort has been directed towards the mechanisms of myocardial dysfunction. The use of bedside techniques to measure ventricular volumes and ejection fractions has increased our insight into myocardial behaviour. It has been appreciated that different mechanisms may result in decreased myocardial function. Furthermore, reversibility of myocardial dysfunction has been documented. And, recently, the role of peripheral vascular dysfunction has come into focus. The complex interactions between cardiac and peripheral vascular abnormalities result in inadequate and maldistributed nutritional flow, progressive organ dysfunction and multiple organ failure. The ultimate goal is to find reversible mechanisms and therapeutic modalities for improving outcome for patients with severe septic shock.

References

Abraham E, Bland RD, Cobo JC, Shoemaker WC (1984) Sequential cardiorespiratory patterns associated with outcome in septic shock. Chest 85:75–80

Baumgartner JD, Vassey C, Perret C (1984) An extreme form of the hyperdynamic syndrome of septic shock. Intensive Care Med 10:245–249

Bell H, Thal A (1970) The peculiar hemodynamics of septic shock. Postgrad Med J 46:106–114

Blain CM, Anderson TO, Pietras RJ, Gunnar RM (1970) Immediate hemodynamic effects of gram-negative vs gram-positive bacteremia in man. Arch Intern Med 126:260–265

Bronsveld W, van Lambalgen AA, van Velzen D, van den Bos GC, Koopman PAR, Thijs LG (1985) Myocardial metabolic and morphometric changes during canine endotoxin shock before and after glucose–insulin–potassium. Cardiovasc Res 19:455–464

Cain SM (1984) Supply dependency of oxygen uptake in ARDS: myth or reality? Am J Med Sci 288:119–124

Calvin JE, Driedger AA, Sibbald WJ (1981a) The hemodynamic effect of rapid fluid infusion in critically ill patients. Surgery 90:61–76

Calvin JE, Driedger AA, Sibbald WJ (1981b) An assessment of myocardial function in human sepsis utilizing ECG gated cardiac scintigraphy. Chest 80:579–586

Cerra FB, Hassett J, Siegel JH (1978) Vasodilator therapy in clinical sepsis with low output syndrome. J Surg Res 25:180–183

Clowes GHA, Farrington GH, Zuschneid W, Cossette GR, Saravis C (1970) Circulating factors in the etiology of pulmonary insufficiency and right heart failure accompanying severe sepsis (peritonitis). Ann Surg 171:663–678

Clowes GHA, Hirsch E, Williams L, Kwasnik E, O'Donnell TF, Cuevas P, Saini VK, Moradi I, Farizan M, Saravis C, Stone M, Kuffler J (1975). Septic lung and shock lung in man. Ann Surg 181:681–692

Clowes GHA, Martin H, Waiji S, Hirsch E, Gazitua R, Goodfellow R (1978) Blood insulin responses to blood glucose levels in high output sepsis and septic shock. Am J Surg 135:577–583

Cohn JD, Greenspan M, Goldstein CR, Gudwin AL, Siegel JH, Del Guercio RM (1968) Arteriovenous shunting in high cardiac output shock syndromes. Surg Gynecol Obstet 127:282–288

Cronenwett JL, Lindenauer SM (1979) Direct measurement of arteriovenous anastomotic blood flow in the septic canine hindlimb. Surgery 85:275–282

Ellrodt AG, Riedinger MS, Kimchi A, Berman DS, Maddahi J, Swan HJC, Murata GH (1985) Left ventricular performance in septic shock: reversible segmental and global abnormalities. Am Heart J 110:402–409

Finley RJ, Duff JH, Holliday RL, Jones D, Marchuk JB (1975) Capillary muscle blood flow in human sepsis. Surgery 78:87–94

Goldfarb RD (1982) Cardiac mechanical performance in circulatory shock: a critical review of methods and results. Circ Shock 9:633–653

Groeneveld ABJ, Bronsveld W, Thijs LG (1986) Hemodynamic determinants of mortality in human septic shock. Surgery 99:140–153

Gunnar RM, Loeb HS, Winslow EJ, Blain C, Robinson J (1973) Hemodynamic measurements in bacteremia and septic shock in man. J Infect Dis 128:S295–S298

Hess ML, Hastillo A, Greenfield LJ (1981) Spectrum of cardiovascular function during gram-negative sepsis. Progr Cardiovasc Dis 23:279–298

Hinshaw LB, Archer LT, Spitzer JJ, Black MR, Peyton MD, Greenfield LJ (1974) Effects of coronary hypotension and endotoxin on myocardial performance. Am J Physiol 227:1051–1057

Hoffman MJ, Greenfield LJ, Sugerman HJ, Tatum JL (1983) Unsuspected right ventricular dysfunction in shock and sepsis. Ann Surg 198:307–319

Houtchens BA, Westenskow DR (1984) Oxygen consumption in septic shock. Circ Shock 13:361–384

Kimchi A, Ellrodt AG, Berman DS, Riedinger MS, Swan HJC, Murata GH (1984) Right ventricular performance in septic shock: a combined radionuclide and hemodynamic study. J Am Coll Cardiol 4:945–951

Krausz MM, Perel A, Eimerl D, Cotev S (1977) Cardiopulmonary effects of volume loading in patients in septic shock. Ann Surg 185:429–434

Kreger BE, Craven DE, McGabe WR (1980) Gram-negative bacteremia: re-evaluation of clinical features and treatment in 612 patients. Am J Med 68:344–355

Kwaan HM, Weil MH (1969) Differences in the mechanism of shock caused by bacterial infections. Surg Gynecol Obstet 128:37–45

Lambalgen AA van, Bronsveld W, van den Bos GC, Thijs LG (1984) Distribution of cardiac output, oxygen consumption and lactate production in canine endotoxin shock. Cardiovasc Res 18:195–205

Lefer AM (1979) Mechanisms of cardiodepression in endotoxin shock. Circ Shock [Suppl.] 1:1–8

Maksad AK, Cha CJ, Stuart RC, Brosco FA, Clowes GHA (1979) Myocardial depression in septic shock: physiologic and metabolic effects of a plasma factor on an isolated heart. Circ Shock [Suppl.] 1:35–42

McGabe WR (1974) Gram-negative bacteremia. Adv Intern Med 19:135–158

Nespoli A, Chiara O, Clement MG, Dagnino G, Bevilacqua G, Aguggini G (1983) The cardiorespiratory impairment in cirrhosis and sepsis. An experimental interpretation using octopamine infusion. Circ Shock 10:15–30

Nishijma H, Weil MH, Shubin H, Cavanilles J (1973) Hemodynamic and metabolic studies on shock associated with gram-negative bacteremia. Medicine 5:287–294

O'Donnell TF, Clowes HA, Talamo RC, Colman RW (1976) Kinin activation in the blood of patients with sepsis. Surg Gynecol Obstet 143:539–545

Parillo JE (1985) Cardiovascular dysfunction in septic shock: new insights into a deadly disease. Int J Cardiol 7:314–321

Parker JL, Adams HR (1985) Isolated cardiac preparations: models of intrinsic myocardial dysfunction in circulatory shock. Circ Shock 15:227–245

Parker MM, Shelhamer JH, Natans MC, Miller L, Masur H, Parillo JE (1983) Serial hemodynamic patterns in survivors and non-survivors of septic shock in humans. Clin Res 31:671A (abstr)

Parker MM, Shelhamer JH, Bacharach SL, Green MV, Natanson C, Frederick TM, Damske BA, Parillo JE (1984) Profound but reversible myocardial depression in patients with septic shock. Ann Intern Med 100:483–490

Postel J, Schloerb PR (1977) Cardiac depression in bacteremia. Ann Surg 176:74–82

Reines HD, Halushka PV, Cook JA, Wise WC, Rambo W (1982) Plasma thromboxane concentrations are raised in patients dying with septic shock. Lancet II:174–175

Robinson JA, Klodnycky ML, Loeb HS, Racic MR, Gunnar RM (1975) Endotoxin, prekallikrein, complement and systemic vascular resistance. Am J Med 59:61–67

Schneider AJ, Teule GJJ, Kester ADM, Heidendal GAK, Thijs LG (1986) Biventricular function during volume loading in porcine E. coli septic shock, with emphasis on right ventricular function. Circ Shock 18:53–63

Shoemaker WC (1971) Cardiorespiratory patterns in complicated and uncomplicated septic shock. Ann Surg 174:119–125

Sibbald WJ, Paterson NAM, Holliday RL, Anderson RA, Lobb TR, Duff JH (1978) Pulmonary hypertension in sepsis. Measurement by the pulmonary arterial diastolic–pulmonary wedge pressure gradient and the influence of passive and active factors. Chest 73:583–591

Sibbald W, Peters S, Lindsay RM (1980) Serotonin and pulmonary hypertension in human septic ARDS. Crit Care Med 8:490–494

Siegel JH, Greenspan M, Del Guercio LRM (1967) Abnormal vascular tone, defective oxygen transport and myocardial failure in human septic shock. Ann Surg 165:504–517

Siegel JH, Cerra FB, Coleman B, Giovannini I, Shetye M, Border JM, McMenamy RH (1979) Physiological and metabolic correlation in human sepsis. Surgery 86:163–193

Siegel JH, Giovannini I, Cerra FB, Nespoli A (1982) Pathologic synergy in cardiovascular and respiratory compensation with cirrhosis and sepsis. A manifestation of a common metabolic defect? Arch Surg 117:225–238

Teule GJJ, van Lingen A, Verweij-van Vught MAAJ, Kester ADM, Mackaay RCM, Bezemer PD, Heidendal GAK, Thijs LG (1984) Role of peripheral pooling in porcine *Escherichia coli* sepsis. Circ Shock 12:115–123

Teule GJJ, van Lingen A, Schneidger AJ, Verweij-van Vught MAAJ, Kester ADM, Heidendal GAK, Thijs LG (1985) Left and right ventricular function in porcine *Escherichia coli* sepsis. Circ Shock 15:185–192

Thijs LG, Teule GJJ, Bronsveld W (1984) Problems in the treatment of septic shock. Resuscitation 11:147–155

Udhoji VN, Weil MH (1965) Hemodynamic and metabolic studies on shock associated with bacteremia. Ann Intern Med 62:966–978

Vincent JL, Weil MH, Puri V, Carlson RW (1981) Circulatory shock associated with purulent peritonitis. Am J Surg 142:262–270

Weil MH, Nisjima H (1978) Cardiac output in bacterial shock. Am J Med 64:920–922

Weisel RD, Vito L, Dennis RC, Valeri CR, Hechtman HB (1977) Myocardial depression during sepsis. Am J Surg 133:511–521

Wiles JB, Cerra FB, Siegel JH, Border JR (1980) The systemic septic response: does the organism matter? Crit Care Med 82:55–60

Wilson RF, Thal AP, Kindling PH, Grifka T, Ackerman E (1965) Hemodynamic measurements in septic shock. Arch Surg 91:121–129

Wilson RF, Chiscano AD, Quadros E, Tarver M (1967) Some observations on 132 patients with septic shock. Anesth Analg 46:751–763

Wilson RF, Sarver EJ, LeBlanc PL (1971) Factors affecting hemodynamics in clinical shock with sepsis. Ann Surg 174:939–943

Winslow EJ, Loeb HS, Rahimtoola SH, Kamath S, Gunnar RM (1973) Hemodynamic studies and results of therapy in 50 patients with bacteremic shock. Am J Med 54:421–432

Chapter 7

Oxygen Delivery and Consumption in the Critically Ill: Their Relation to the Development of Multiple Organ Failure

D. Bihari

Despite the many recent advances in intensive therapy, the development of multiple organ dysfunction associated with acute respiratory failure occurring as a consequence of septic shock or severe established sepsis continues to be associated with a poor prognosis (Ledingham et al. 1982; Bell et al. 1983; Andreadis and Petty 1985; Petty 1985; Seidenfeld et al. 1986; anonymous 1986a). Similarly, patients who suffer an episode of massive trauma with widespread tissue necrosis are also at risk of developing this syndrome with its attendant high mortality—even in the absence of overt sepsis (Baue 1975; Borzotta and Polk 1983; Faist et al. 1983).

Whilst it is generally accepted that it is possible to resuscitate adequately the majority of patients with septic and traumatic shock using the conventional techniques of fluid replacement (assessed by invasive haemodynamic monitoring) and mechanical ventilation, it is more difficult to avoid the subsequent development of acute renal and respiratory failure (Wardle 1982; Cameron 1986), hepatic dysfunction (Cerra et al. 1979; Nolan 1981; Chaudry et al. 1986; Gimson 1987), stress ulceration of the gastrointestinal tract (Knight et al. 1985), encephalopathy (Hasselgren and Fischer 1986), coagulation disturbances, acute heart failure (Parker et al. 1984a; Sibbald and Driedger 1986) and other less well-defined system failures which constitute the multiple organ failure syndrome (Table 7.1). This is of great importance, for many investigators have emphasized that mortality following an episode of septic shock or multiple trauma is directly related to the number of organs that appear to have failed (Tilney et al. 1973; Fry et al. 1980; Knaus et al. 1985). Thus, the prevention of multiple organ failure assumes an important priority in the care of the critically ill.

Table 7.1.   The multiple organ failure syndrome associated with sepsis and trauma

---

1.   Acute respiratory failure, "ARDS"
2.   Acute renal failure, "ATN" (commonly associated with the administration of nephrotoxins, e.g. aminoglycosides and non-steroidal anti-inflammatory drugs)
3.   Stress ulceration of the gastrointestinal tract
4.   Acute hepatic dysfunction
5.   Acalculous cholecystitis
6.   Acute pancreatitis
7.   Encephalopathy
8.   Acute heart failure (frequently right ventricular dysfunction as a result of pulmonary hypertension)
9.   Disturbances in coagulation and fibrinolysis
10. Defective reticuloendothelial cell function
11. Immune deficiency state
12. Endocrine abnormalities (e.g. relative cortisol deficiency, reductions in plasma ionized calcium, abnormalities in thyroid and pituitary function)
13. Skeletal muscle abnormalities (wasting and weakness)
14. Other, less well-defined organ systems failure

Possible Mechanisms Contributing to the Development of Multiple Organ Failure

There are a number of hypotheses concerning the pathogenesis of this multiple organ dysfunction associated with septic and traumatic states (George and Tinker 1983), but in general these speculative theories (which are not necessarily mutually exclusive) may be divided into three main groups (Bihari et al. 1986b; Coalson 1986) (Table 7.2).

The first assumes a *maldistribution of blood flow* within the microcirculations of respiring tissues either as a result of microembolic phenomena (Saldeen 1976; Scovill et al. 1978; Shah et al. 1981; Smith et al. 1981; Malik 1983; Saba 1986; Till and Ward 1986) a consequence of complement activation with the formation of leucocyte/ platelet aggregates, local activation of the clotting cascade with fibrin deposition, accumulation of particulate matter secondary to reticuloendothelial cell dysfunction related in part to a plasma fibronectin deficiency, and/or resulting from the release of various vasoactive compounds such as endotoxin (Morrison and Ulevitch 1978; Brigham and Meyrick 1986), opioids (Bernton et al. 1985), histamine, bradykinin, serotonin (Bond and Johnson 1985) and various eicosanoids (Lefer 1985).

Secondly, there are those that emphasize *the inappropriate and uncontrolled release of cytotoxic mediators*: from polymorphonuclear leucocytes (anonymous 1986a, b; Warshawski et al. 1986; Weiland et al. 1986), in the form of oxygen free radicals and proteases (Bertrand 1985; Westaby 1986) or metabolites of arachidonic acid (Brigham 1985; Gee et al. 1985); from stimulated monocytes and fixed tissue macrophages which also secrete various different monokines (such as interleukin 1, tumour necrosis factor and proteolysis inducing factor) mediating the acute phase response (Baracos et al. 1983; Clowes et al. 1983; Filkins 1985; Old 1985); and from platelets (which primarily release thromboxane A_2) (Smith et al. 1981; Lefer 1985).

Finally some investigators have emphasized *a specific defect in cellular metabolic activity* as a direct consequence of a (some) circulating "toxic" factor(s), such as endotoxin itself, which might in some way inhibit mitochondrial respiratory function (Schumer et al. 1970; Mela et al. 1971; Mela 1983; Houtchens and Westenskow 1984). This latter view has been supported by the finding of a reduced total body oxygen consumption (VO_2) in some patients with sepsis and trauma suggesting the development of a blockage in the utilization of oxygen by tissues (Siegel et al. 1974, 1981; Cerra et al. 1979)—so-called histiotoxic hypoxia (Barcroft 1920).

In the past it has been relatively difficult to investigate these differing hypotheses directly other than in the various animal models of shock, studies in patients having been limited to the measurement of the plasma concentration of a single or some proposed group of mediator followed by an attempt to correlate the results with organ dysfunction. Moreover, many clinical investigations have centred primarily on the relatively easy-to-assess cardiopulmonary dysfunction rather than on the more inaccessible but widespread disturbances in microcirculatory blood flow and tissue oxygenation. Nevertheless, an assessment of the relative importance of the different mediator systems in the pathogenesis of tissue injury remains an extremely important aspect of shock research and is dealt with elsewhere in this volume. Only on this basis will rational pharmacological interventions in the form of "shock cocktails" become available in the future in an effort to inhibit the activation of the various cellular and cascade systems and to block the synthesis and effects of the different mediators which undoubtedly contribute to tissue damage.

Table 7.2. Possible mechanisms contributing to the development of multiple organ failure associated with sepsis and trauma

---

1. Mechanical factors producing a maldistribution of blood flow within the microcirculation leading to hypoxic damage:

    (a) Microembolic phenomena
        Rigid activated leucocytes
        Decreased red cell deformability
        Leucocyte and platelet emboli
        Localized disseminated intravascular coagulation, fibrin deposition
        Accumulation of particulate matter (secondary in part to defective reticuloendothelial function)
        Increased capillary permeability leading to interstitial oedema and capillary compression

    (b) Vasoactive compounds disturbing normal autoregulation
        Endotoxin
        Catecholamines
        Opioids, serotonin, bradykinin
        Eicosanoids (especially TXA_2 and the leucotrienes LTB4, LTC4 and LTD4)

2. Inappropriate and uncontrolled release of cytotoxic mediators directly damaging membranes and cells.

    (a) Oxygen free radicals, proteases, lysosomal enzymes and various eicosanoids from complement-activated neutrophils and stimulated monocytes/macrophages

    (b) Various monokines including interleukin 1, tumour necrosis factor and proteolysis inducing factor from stimulated monocytes and macrophages

    (c) Thromboxane A_2 from activated platelets

3. ? Specific defect in cellular and mitochondrial oxidative capacity as a consequence of direct inhibition, possibly by endotoxin. Probably only of importance in terminal and irreversible stages of shock

---

Tissue Hypoxia in the Critically Ill

Given the complexity surrounding the interactions of the mediator systems (and the relative failure of available drugs, such as steroids and cyclo-oxygenase inhibitors, used in isolation, to make much impact upon outcome in patients) the clinician is left with one absolute therapeutic principle: to prevent organ dysfunction as a direct consequence of inadequate tissue perfusion. This is easier said than done, especially as severe tissue hypoxia may occur in the presence of an apparently adequate systemic blood flow, pressure and oxygenation as a consequence of abnormalities in the distribution of blood flow within the microcirculation (Miller 1982; Bihari and Tinker 1983; Schumacker and Wood 1984; Bihari et al. 1986c; Shoemaker 1986).

Since the introduction of relatively simple and reliable assays for plasma and whole blood lactate, many investigators have emphasized the undeniable importance of the presence of a lactic acidosis in prognosis from shock (Cohen and Woods 1976; Ledingham et al. 1982; Park and Arieff 1983). The conventional view holds that an increase in blood lactate concentration or in the lactate:pyruvate ratio is good evidence of critical tissue hypoxia (Cryan and Ledingham 1986). Nevertheless, there may be alternative explanations for a raised blood lactate in the critically ill other than anaerobic respiration—e.g. a respiratory alkalosis with increased lactate production (Berry and Scheuer 1967) or reductions in liver blood flow and hepatic clearance (Burns et al. 1982; Cohen and Woods 1983; Cowan et al. 1984; Bihari et al. 1985a). Ozawa et al. (1983) have suggested that disturbances in the ketone body ratio may be a more sensitive reflection of hepatic mitochondrial redox potential and the probability of postoperative organ failure (anonymous 1984) whilst there may be increases in the mixed venous concentrations of the various purine nucleotide degradation products of ATP in critically ill patients in the absence of any other suggestion of tissue hypoxia (Grum et al. 1985). Nevertheless, there are good biochemical reasons for believing that critical tissue hypoxia with defective organ function can occur not only in the absence of hyperlactataemia but also long before the presumed switch from aerobic to anaerobic respiration within cells (Robin 1980; Denison 1981). It is important to remember that under normal resting conditions some 10%–20% of oxygen consumed by the body is used not in oxidative phosphorylation but in other extramitochondrial oxygen-requiring reactions. These reactions, which are catalysed by enzymes with a much lower affinity for oxygen, i.e. with an increased Michaelis constant (Fig. 7.1), may be inhibited by hypoxia at cellular oxygen tensions well above the Pasteur point. Therefore, an extramitochondrial oxygen debt might occur early on in the course of critical illness in the absence of a lactic acidosis (thus making the blood lactate concentration less useful in diagnosis) and contribute to the development of multiple organ dysfunction and subsequent failure.

The widespread introduction of pulmonary artery catheterization with the measurement of cardiac output by thermodilution has enabled a full characterization of the wide spectrum of haemodynamic abnormalities occurring in sepsis and trauma (Parker et al. 1984a, b). It is true to say that the cardiac output may be low, normal or high depending upon the phase of the illness, the volaemic status of the patient and the presence or absence of coexisting coronary artery disease. However, more importantly, the various circulatory disturbances have been shown to result in marked abnormalities of systemic oxygen transport and utilization such that in contrast with the normal control of oxygen metabolism *total body oxygen consumption*

Michaelis–Menten equation for O_2-consuming enzymes:

$$\dot{V}_{O_2}/\dot{V}_{O_2}\ \text{max} = P_{O_2}/(P_{O_2} + KmO_2)$$

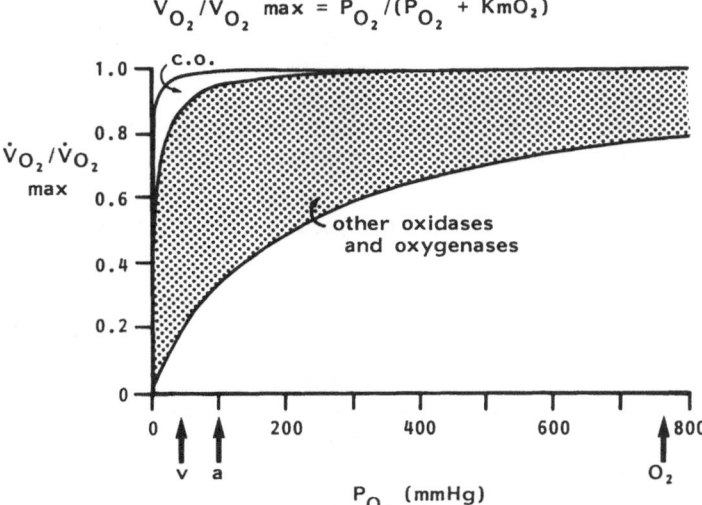

Fig. 7.1. A plot of oxygen consumption ($\dot{V}O_2$) against oxygen concentration (PO_2) for three hypothetical reactions with Michaelis constants for oxygen (KmO_2) of 0.5, 1.5 and 200 mmHg, obeying the Michaelis–Menten Law. c.o., cytochrome oxidase a_3; v, a and O_2, the venous, arterial and pure oxygen values for oxygen tension. (Redrawn from Denison (1981), with permission.)

becomes dependent upon the convection of oxygen from the lungs to capillaries, even at very high flows.

This particular abnormality suggests that previous observations of a low oxygen consumption in sepsis (Siegel et al. 1967; Duff et al. 1969; Houtchens and Westenskow 1984) were most likely related to a primary reduction in cardiac output, a consequence of both relative hypovolaemia occurring as a result of inappropriate vasodilation and direct myocardial depression associated with severe sepsis. Other than some in vitro studies of mitochondrial respiration with very high doses of endotoxin (Mela 1983) there is very little evidence that a septic illness (other than immediately pre-mortem) is associated with a primary reduction in oxygen consumption—indeed most clinical investigations have concluded that this hypercatabolic state requires an unpredictable increase in oxygen consumption (Bartlett et al. 1982; Carlsson et al. 1984; Mann et al. 1985; Weissman et al. 1986) which occurs provided oxygen is available.

Oxygen Consumption in the Critically Ill

Given that transport of substrates to the tissues and removal of the waste products of metabolism from the tissues are the major functions of the circulation and that oxygen is the most flow dependent substrate as it has the greatest extraction ratio, then it follows that the adequacy of tissue perfusion is best evaluated in terms of oxygen delivery to tissues and oxygen consumption (Shoemaker 1986). Shoemaker has popularized the concept of an "arterial oxygen delivery" (DaO_2), i.e the total quantity of oxygen transported towards the tissues in any one minute (Shoemaker and Czer 1979). This measure of oxygen transport is derived from the product of the cardiac index (CI) (measured by thermodilution) and the arterial oxygen content (CaO_2) (most often calculated from direct measurements of haemoglobin with percentage saturation [oximetry] and the oxygen tension in solution) and is expressed in millilitres of oxygen delivered per minute:

$$DaO_2 \text{ (ml } O_2 \text{ min}^{-1} \text{ m}^{-2}) = CI \text{ (l min}^{-1} \text{ m}^{-2}) \times CaO_2 \text{ (vol \%)} \times 10$$

This derived variable is a useful guide to systemic oxygenation but is unfortunately misnamed for it only describes the "despatch" of oxygen from the left ventricle towards the periphery rather than "delivery" (with the connotation of availability) of oxygen within the microcirculation of actively respiring tissues. This distinction between "despatch" and "delivery" may seem pedantic but assumes considerable importance in understanding the development of tissue hypoxia in the face of a normal or increased DaO_2. Indeed, it is the change in distribution of the oxygen supply— or, to put it another way, an increase in the normal "heterogeneity" of blood flow within the microcirculation *unrelated* to the oxygen demand of respiring tissues—that underlies the apparent peripheral arteriovenous "shunting" of sepsis (Siegel et al. 1967; Cohn et al. 1968; Carroll and Snyder 1980), the development of a tissue oxygen debt and finally defective organ function.

The Normal Relationship Between DaO_2 and Consumption

There is now a large body of evidence available from animal studies suggesting that under normal physiological conditions *oxygen consumption is independent of blood flow and DaO_2 above some critically low value* (Cain 1977; Shah et al. 1981; Adams et al. 1982). Thus, above this critical oxygen delivery there is no relation between delivery and consumption whereas below the critical value there is a close correlation between the two, consumption apparently being limited by delivery. Similarly, in normal anaesthetized man, oxygen consumption appears to be independent of delivery above the critical DaO_2 of 330 ml O_2 min^{-1} m^{-2}, so that changes in DaO_2 above this level are not associated with changes in consumption (Shibutani et al. 1983), the level of oxygen uptake presumably being set by cellular metabolic demands (Grainger et al. 1975). Of course, this description of the physiological relation, which has been confirmed in patients with chronic cardiopulmonary disease (Daniel et al. 1978; Rubin et al. 1982; Chappell et al. 1983; Brent et al. 1984), is only accurate in the presence of a stable oxygen consumption which is not changing as a result of some primary change in metabolic rate such as is produced by a change in temperature, activity of the study subject (Weissman et al. 1984) or concentration of catecholamines or other stress hormones (Askanazi et al. 1986). In this setting, one would expect changes in DaO_2 to be related to changes in oxygen consumption mediated through the processes of normal autoregulation.

The normal resting DaO_2 in man in our and other laboratories is 500–700 ml O_2 min^{-1} m^{-2}, indicating a substantial reserve in the oxygen supply. This may be conceived of as "shunted oxygen"—i.e. available oxygen that passes through the microcirculation unused. In the presence of a normal resting oxygen consumption of 120–160 ml O_2 min^{-1} m^{-2}, measured directly or indirectly by the reverse Fick principle, this gives an extraction ratio of 22%–30%.

Abnormalities in the Relation of DaO_2 to Consumption

Powers et al. (1973) were one of the first groups to focus on abnormalities in the relation between oxygen delivery and consumption in critically ill patients with apparently adequate systemic blood pressure, flow and oxygenation; their observations have been confirmed by many others (Rhodes et al. 1978; Danek et al. 1980; Mohsenifar et al. 1983; Kariman and Burns 1985). A close correlation between oxygen consumption and DaO_2 has been observed over a wide range of values above the critical DaO_2 in patients with septic shock and in those with acute respiratory failure associated with sepsis and severe trauma. The suggestion has been that this relationship is highly abnormal and reflects in some way a disturbance in the control of oxygen metabolism, perhaps due to defective autoregulation (Cain 1984; Schumacker and Wood 1984; Schumacker and Cain 1987).

However, there are a number of problems with these studies in so far as the majority have depended upon repeated measurements of oxygen delivery and uptake over a period of days in an intensive therapy unit, during which time the metabolic rate of the patients studied was not necessarily constant. Again, it is important to emphasize that the normal physiological relation between delivery and consumption is such that as metabolic demand changes so will the supply of oxygen. This is particularly evident in aerobic exercise, during which the cardiac output and the oxygen extraction ratio increase to maintain the oxygen supply in the face of an increased demand. Therefore, it is not surprising that there is a statistical correlation between delivery and consumption in the critically ill when these two variables are measured over a prolonged period of time during which oxygen demand is most probably changing. Furthermore, the maintenance of this relationship suggests an intact mechanism of autoregulation of blood flow so that changes in oxygen demand in various tissues are indeed met by appropriate changes in distribution and supply.

More importantly, some investigators have demonstrated that in patients with acute respiratory failure secondary to sepsis or trauma and requiring mechanical ventilation, an acute reduction of DaO_2 (achieved by dropping the cardiac output by the application of positive end-expiratory pressure (PEEP)) is accompanied by a fall in oxygen uptake (Powers et al. 1973; Danek et al. 1980). This finding suggests that in these patients in whom the demand for oxygen during the study presumably did not change, there was indeed some disturbance of autoregulation so that as DaO_2 fell as a consequence of PEEP, oxygen uptake was reduced, limited by delivery, and this may have led to the development of a tissue oxygen debt. It is important to emphasize that in these studies all patients had a DaO_2 before and after PEEP substantially greater than that thought to limit oxygen uptake under normal physiological conditions. Nevertheless, intervention in the form of PEEP may cause some primary maldistribution of blood flow (Long et al. 1984) and does not assess the effect of increasing oxygen despatch to tissues. Thus, one is unable from these investigations to comment upon the possibility that the patients studied had an undetected oxygen debt or a disturbance in autoregulation before the application of PEEP.

Measurement of Oxygen Consumption in the Critically Ill

Another criticism of these studies has concerned the various methods used to measure oxygen consumption in the critically ill (Nelson et al. 1982; Pepe and Culver 1982; Davila 1984; Schumacker and Wood 1984). Many clinical investigators have relied upon the calculation of oxygen consumption (i.e. the oxygen uptake index, OUI) by the reverse Fick principle:

$$\text{OUI (ml O}_2 \text{ min}^{-1} \text{ m}^{-2}) = \text{cardiac index (CI)} \times 10$$
$$\times \text{(arteriovenous O}_2 \text{ content difference)}$$

It has not been clear whether it is statistically correct or appropriate to relate DaO_2 to OUI, as the two variables are dependent at source, i.e. on the measurement of cardiac output.

We have studied this problem by obtaining direct and indirect measurements of OUI in 2 human volunteers before and during the infusion of the microcirculatory vasodilator prostacyclin (PGI_2) and comparing this normal delivery/consumption relationship with that obtained in 2 postoperative patients studied over a more prolonged period of time (Bihari et al. 1987). We measured OUI directly using a microprocessor controlled metabolic cart (Beckman MMC Horizons Systems) and derived CI by the Fick equation. These values were then compared with thermodilution measurements of CI and the calculated reverse Fick OUI.

Not surprisingly, given that the two methods are supposed to measure the same physiological variables (Bland and Altman 1986), we, as others have done before (Rubin et al. 1982; Chappell et al. 1983), obtained close correlations between both Fick and thermodilution CI and direct and indirect OUI (Fig. 7.2a, b). The correlation was closer for CI because the range of measured values was wider (cardiac output having been substantially increased in our human volunteers by vasodilatation), but even so there was a very wide *range of agreement* (Fig. 7.3a, b). This was especially obvious for measurements of OUI, in which the 95% confidence limits for the agreement between the two methods suggested that one might differ from the other by as much as 70 ml O_2 min^{-1} m^{-2} in one measurement in 20. Nevertheless, the relations between DaO_2 and OUI, however measured, were the same in the subjects studied (Fig. 7.4a, b). Furthermore, an assessment of the repeatability of the measurements of OUI, direct and indirect, was made by obtaining the coefficient of variation for the two methods (Fig. 7.5). Although the mean coefficient of variation (COV) was significantly lower for measurements of OUI using the Beckman metabolic cart compared with the reverse Fick method (mean COV $\pm$ SD of 3.3 ± 2.0 versus 5.3 ± 2.9 respectively, $p < 0.05$), both methods were highly reproducible.

We have interpreted these observations as evidence that the calculated reverse Fick OUI can be used in the place of a directly measured value and related to DaO_2 although the numerical values of directly and indirectly measured OUI may be very different. Other workers have also obtained similar findings, suggesting that it is neither inappropriate nor incorrect to relate changes in DaO_2 (ax) to changes in the reverse Fick OUI (bx) (Cain 1984).

Our study also demonstrated that in human volunteers who were acutely vasodilated with PGI_2 and who subsequently had substantial increases in DaO_2, there was no concomitant increase in OUI (however measured). We have confirmed this finding in 5 other, reasonably fit, conscious patients (with no evidence of multiple organ dysfunction) who required pulmonary artery catheterization for other reasons (3 for

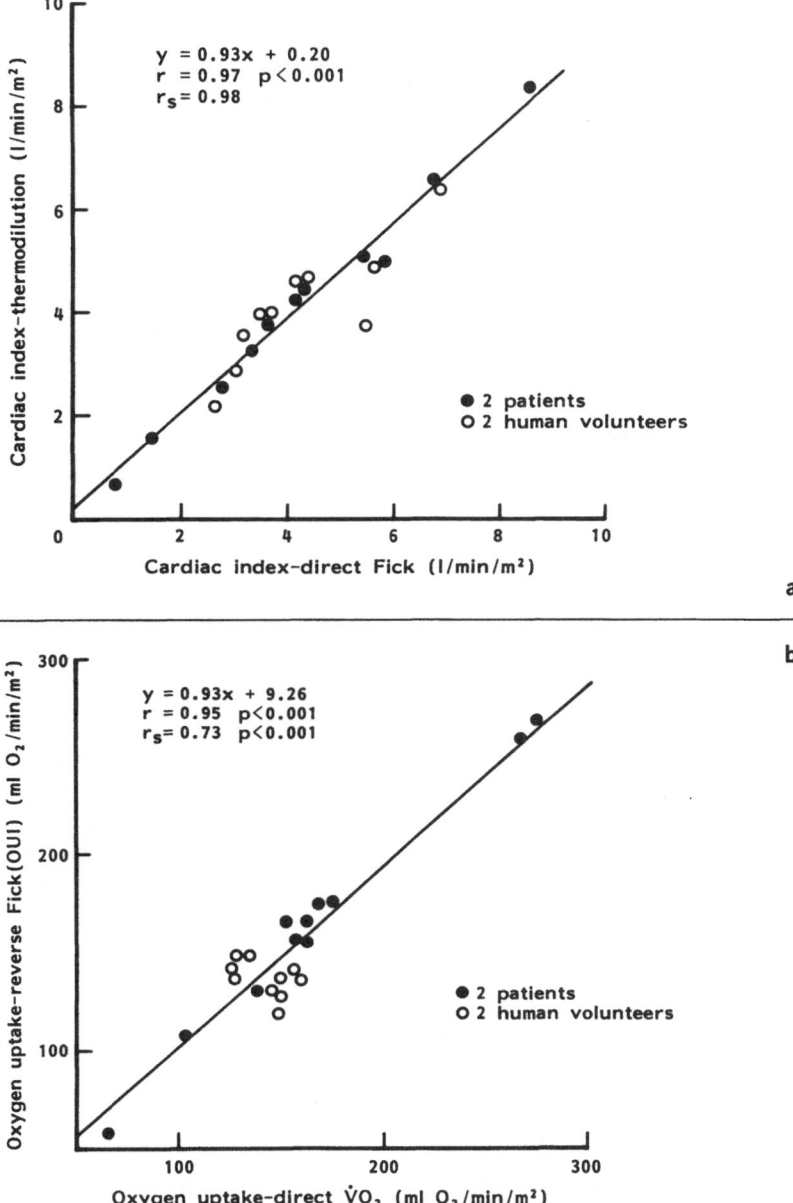

Fig. 7.2. **a** The relation of the cardiac index (CI) calculated from the Fick equation (*x*) with that measured by thermodilution (*y*) in 2 postoperative patients (11 measurements over 48 hours) and 2 human volunteers (10 measurements over 90 minutes before and during a 5 ng kg^{-1} min^{-1} PGI$_2$ infusion). **b** The relation of oxygen consumption measured directly ($\dot{V}O_2$) (*x*) with that calculated from the cardiac index and the arteriovenous oxygen content difference (OUI) (*y*) in 2 postoperative patients (11 measurements over 48 hours) and 2 human volunteers (10 measurements over 90 minutes before and during a 5 ng kg^{-1} min^{-1} PGI$_2$ infusion). (*r* = ordinary correlation coefficient; r_s = Spearman rank correlation coefficient.)

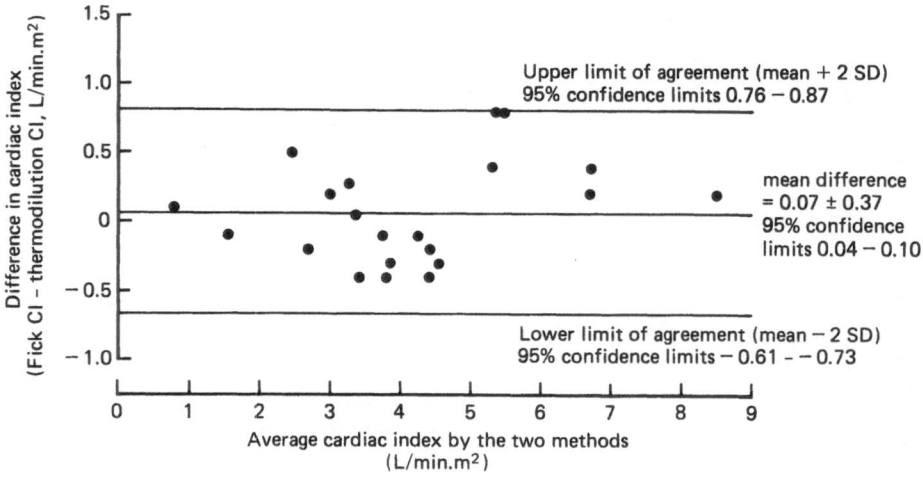

a

b

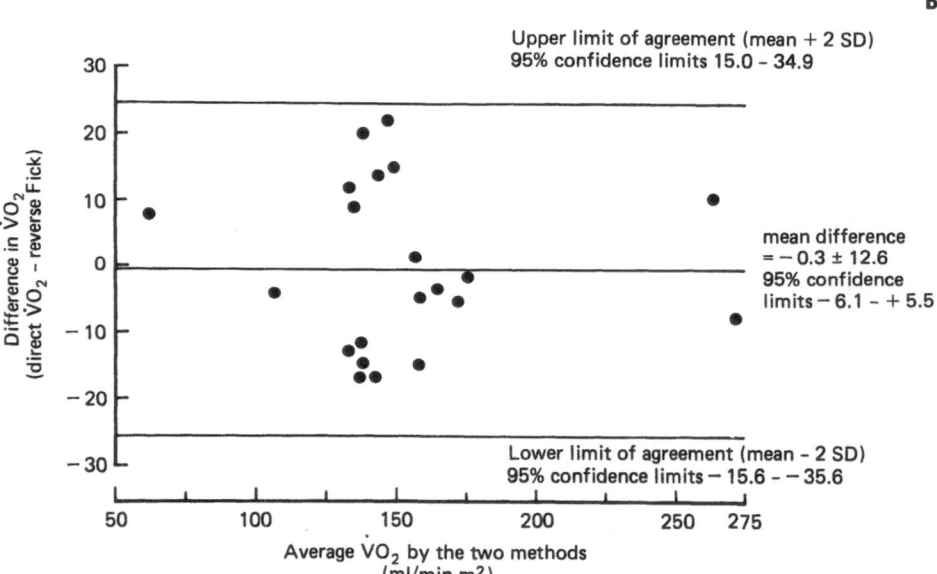

Fig. 7.3. **a** The limits of agreement between the two methods of measuring cardiac index (CI). Although the mean difference between the two methods was only 0.07 ($\pm$ 0.37 sp) l min^{-1} m^{-2}, on the basis of this small set of data the two methods may disagree by as much as 1.6 l min^{-1} m^{-2} in one measurement in 20. **b** The limits of agreement between the two methods of measuring oxygen consumption. Although the mean difference between the two methods was only 0.3 ($\pm$ 12.6 sp) ml O$_2$ min^{-1} m^{-2}, on the basis of this small set of data the two methods may disagree by as much as 70.5 ml O$_2$ min^{-1} m^{-2} in one measurement in 20. This form of analysis demonstrates that correlations alone may not reveal considerable discrepancies between the two methods.

the investigation of idiopathic pulmonary hypertension, 2 for fluid balance studies) (Fig. 7.6). As far as we are aware, this study and that of Nowak and Weunmalm (1978) provide the only control data available on the effects of vasodilatory prostaglandins on DaO_2 and OUI in normal human volunteers—an important consideration in the interpretation of the various studies of PGE_1 and PGI_2 in the management of acute respiratory failure associated with liver failure, sepsis and trauma (Gimson et al. 1984; Tokioka et al. 1985; Holcroft et al. 1986; Shoemaker and Appel 1986). Our results confirm the normal physiological relationship between DaO_2 and OUI— that is the independence of OUI from DaO_2 in this range of delivery—and demonstrate that $PG1_2$ is not a primary metabolic stimulant.

Interpretation of Measurements of Oxygen Consumption

To summarize the discussion thus far, we have concluded that although there are difficulties in measuring oxygen consumption in the critically ill, it is possible to use the calculated reverse Fick OUI which is usually readily available in the intensive therapy unit. Furthermore, it does not seem inappropriate to relate this CI dependent variable to DaO_2, another variable dependent in part upon the measurement of CI. Indeed, this is the only way to interpret measurements of oxygen consumption in so far as a single measurement of either variable on its own has little or no meaning in the assessment of tissue oxygenation. However, provided that some remnants of microcirculatory autoregulation remain intact within normal human volunteers or critically ill patients, one would expect close correlations between OUI and DaO_2 over long periods of time as changes in the demand for oxygen are matched by appropriate changes in supply. It may be that in the more prolonged studies of oxygen consumption in the critically ill it would be more useful to compare the slope of the DaO_2 : OUI relationship in individual patients to obtain some assessment of the efficiency of the oxygen transport system (Fig. 7.7).

Oxygen consumption following trauma or in patients with sepsis can be low, normal or high (Fig. 7.8) depending not only upon the intrinsic metabolic rate of the patient (modulated by many well-defined and as yet unknown determinants) but also upon such limiting factors as transport and distribution of substrate. Traditional teaching emphasizes a short "ebb" phase (with a low oxygen consumption) immediately after injury followed by a more prolonged, reparative "flow" phase (in which the oxygen consumption is raised). In practice, with the introduction of rapid resuscitation and the correction of hypovolaemia, these separate phases have become more difficult to distinguish. The important question is thus no longer whether the patient is in the ebb or flow phase, but rather whether the measured oxygen consumption is adequate for tissue requirements. This is a difficult question to answer (given that an oxygen debt may not necessarily be associated with an elevated blood lactate concentration) but is addressed directly by an "oxygen flux test" (Bihari and Tinker 1983; Bihari and Gimson 1983).

The Oxygen Flux Test

The concept of an oxygen flux test depends upon the normal independence of oxygen consumption from arterial oxygen delivery. Thus, following a set of measurements

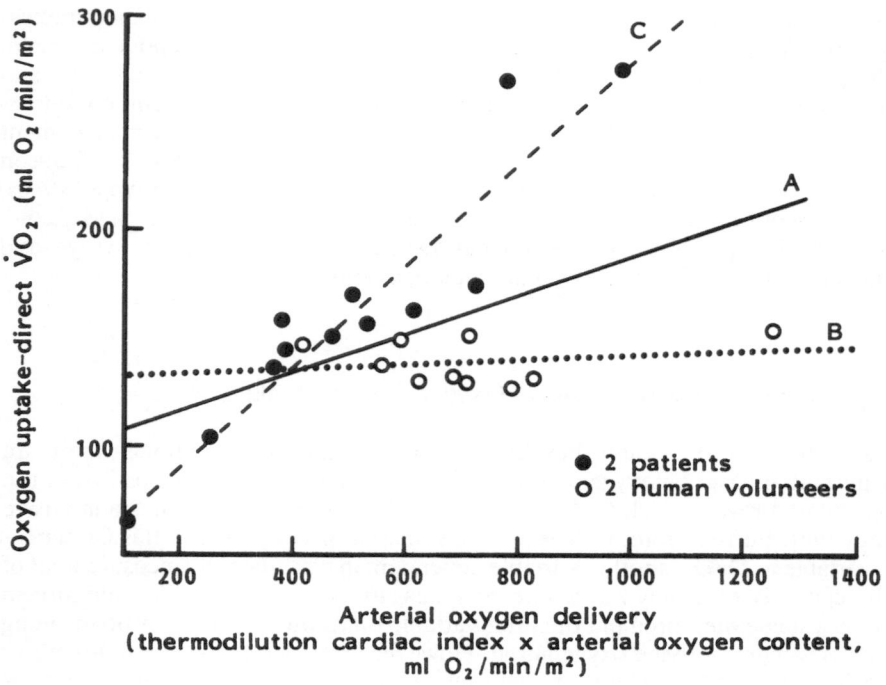

a

b

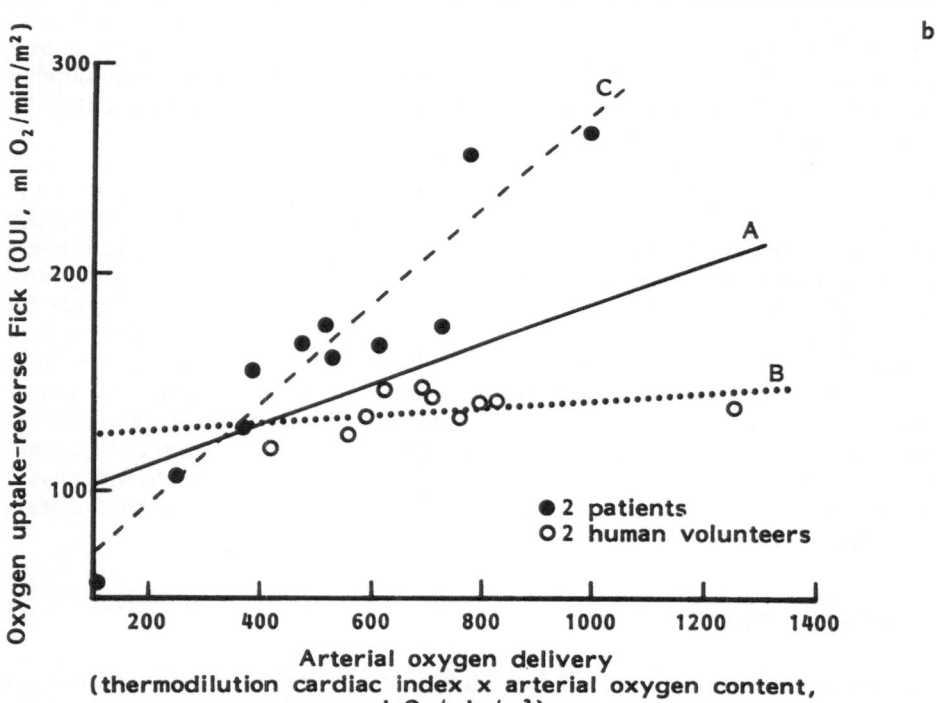

Fig. 7.4. **a** The relation of arterial oxygen delivery (x) to oxygen consumption measured directly ($\dot{V}O_2 = y$) in 2 postoperative patients (11 measurements over 48 hours) and 2 human volunteers (10 measurements over 90 minutes before and during a 5 ng kg^{-1} min^{-1} PGI$_2$ infusion). The regression line A for the 21 sets of data is given by: $y = 0.09x + 96$ ($r = 0.51\ p < 0.02$, $r_s = 0.30\ p$ NS). The regression line B for the 10 sets of data from the 2 human volunteers is given by: $y = 0.01x + 133$ ($r = 0.20\ p$ NS). The regression line C for the 11 sets of data from the 2 patients is given by: $y = 0.23x + 46$ ($r = 0.93\ p < 0.001$, $r_s = 0.94$).

b The relation of arterial oxygen delivery (x) to oxygen consumption calculated from the cardiac index and the arteriovenous oxygen content difference (OUI $= y$) in 2 postoperative patients (11 measurements over 48 hours) and 2 human volunteers (10 measurements over 90 minutes before and during a 5 ng kg^{-1} min^{-1} PGI$_2$ infusion). The regression line A for the 21 sets of data is given by: $y = 0.09x + 95$ ($r = 0.52\ p < 0.02$, $r_s = 0.41\ p$ NS). The regression line B for the 10 sets of data from the 2 human volunteers is given by: $y = 0.02x + 125$ ($r = 0.43\ p$ NS). The regression line C for the 11 sets of data from the 2 patients is given by: $y = 0.23x + 48$ ($r = 0.94\ p < 0.001$, $r_s = 0.94$). ($r =$ ordinary correlation coefficient, $r_s =$ Spearman rank correlation coefficient.)

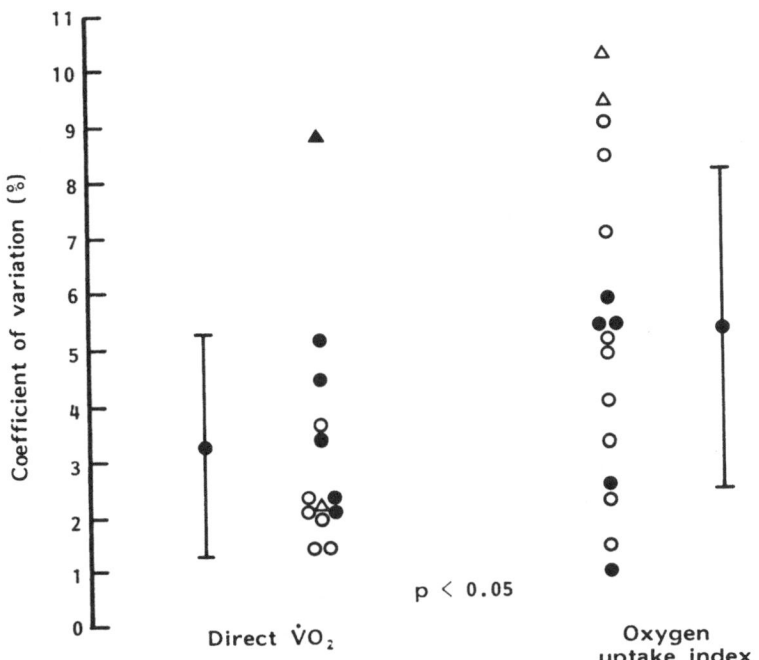

Fig. 7.5. A comparison of the coefficients of variation for the direct and indirect methods of measuring oxygen consumption:
1. Direct $\dot{V}O_2$ (measured using the Beckman MMC Horizon Systems):
   10 measurements were made in 13 periods of 30 minutes in 2 sedated, paralysed postoperative patients receiving mechanical ventilation (FiO$_2$ 0.4–0.45) (O, ●) and two spontaneously breathing human volunteers (△, ▲).
2. Oxygen uptake index (calculated from the reverse Fick relation):
   Three (O), four (△) or five (●) measurements of OUI were made during 16 periods of up to 90 minutes in 13 sedated, paralysed postoperative patients with no evidence of multiple organ dysfunction.

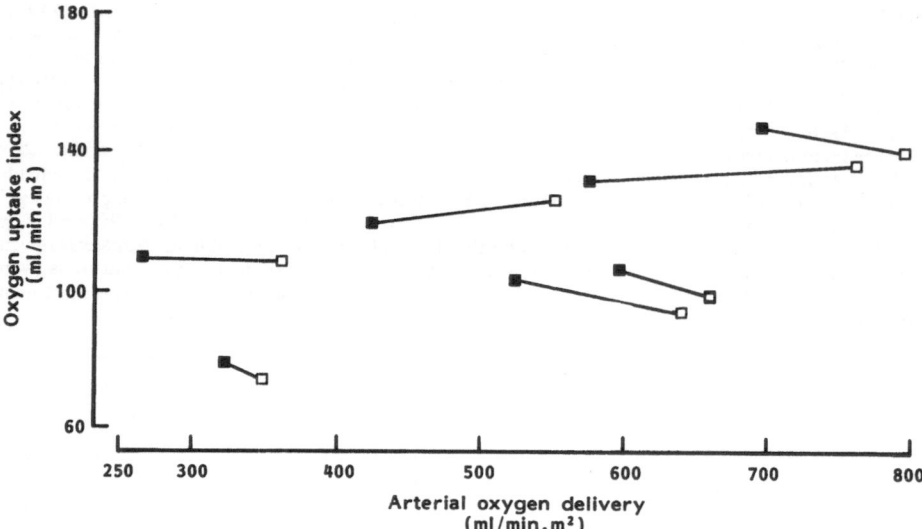

Fig. 7.6. The effect of PGI$_2$ (5 ng kg^{-1} min^{-1}) on arterial oxygen delivery (DaO$_2$) and calculated oxygen uptake index (OUI) in 7 human volunteers with no evidence of multiple organ dysfunction (2 normals, 3 patients undergoing the investigation of idiopathic primary pulmonary hypertension and 2 others undergoing fluid balance studies). Despite substantial increases in DaO$_2$ associated with the vasodilatation produced by the PGI$_2$ infusion (medians before (*filled squares*) and at 30 minutes (*open squares*) 563 and 644 ml min^{-1} m^{-2} respectively, $p < 0.01$; median % change 18%), no significant increase in OUI was observed (median % change in OUI of −3%).

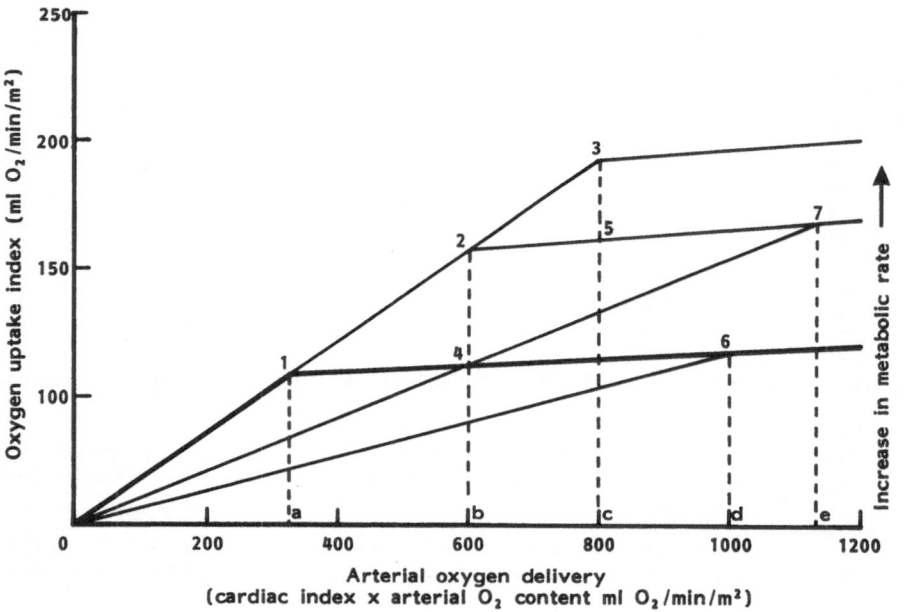

of OUI and DaO_2 (both of which may be in the normal physiological range), DaO_2 is increased within a short period of time (e.g. 30 minutes) during which the metabolic rate of the patient is supposedly constant; the response in terms of a change in OUI is then observed (Fig. 7.9). Two possible results may be obtained: either the OUI remains unchanged despite the higher delivery, suggesting the absence of a tissue oxygen debt; or OUI may increase, the percentage increase presumably depending in part upon the size of any oxygen debt present and the distribution of the extra oxygen delivered. Similarly, the response to a fall in DaO_2 (produced perhaps by the application of PEEP) may be analysed. Provided DaO_2 remains above the physiological critical limit (330 ml O_2 min^{-1} m^{-2} in anaesthetized man) then no change in OUI would be expected. On the other hand, if DaO_2 falls below this critically limiting level or, perhaps more importantly, if defective autoregulation is present such that the smaller quantity of despatched oxygen cannot be appropriately redistributed, then OUI will fall in concert with the reduction in DaO_2. Although this may be a reasonable approach to the identification of "delivery-dependent oxygen consumption", it is unhelpful in identifying an oxygen debt that may have been present already at the higher baseline DaO_2, and it ignores any other effects of PEEP.

We have studied the former approach (i.e. increasing DaO_2) in the diagnosis of tissue hypoxia in two different groups of critically ill patients: patients with severe sepsis and dysfunction of two or more organ systems (Bihari et al. 1986d) and patients with acute liver failure (Gimson et al. 1984; Bihari et al. 1986a, b). This latter group (acute liver failure associated with paracetamol self-poisoning or viral hepatitis) forms an extremely interesting and consistent study group; the patients are young with no pre-existing organic disease and frequently develop multiple organ dysfunction during the course of their illness (Bihari 1985). Invariably, the patients present with a similar but much more profound haemodynamic disturbance than occurs in sepsis (which in our experience has a considerably more varied presentation). We

---

Fig. 7.7. Hypothetical relations between the supply and demand for oxygen in health and disease.

(i) *The resting state.* OUI is usually independent of DaO_2 (*thick line passing through points 1, 4 and 6*) above some critically lower limiting DaO_2 (*point a*), which, in anaesthetized man, is 330 ml O_2 min^{-1} m^{-2}. Nevertheless, there may be a very small increase in OUI (2%–5%) as DaO_2 increases above the critical delivery (which is why the line 1,4,6 is not horizontal); this may be related to the perfusion of relatively ischaemic capillary beds in resting skeletal muscle, an increase in myocardial oxygen consumption required to maintain the higher DaO_2 values and, in those studies in which patients are vasodilated, a compensatory increase in circulating catecholamines.

(ii) *An increase in metabolic rate, e.g. exercise (line 2,5,7).* The critical delivery at which OUI is limited (*point b*), increases as the metabolic rate is elevated. Provided DaO_2 can increase to this point, respiration during exercise (or critical illness) remains aerobic. The *line 1,2,3* and its slope represents the efficiency of the oxygen transport system and is measured by the oxygen extraction ratio (OUI/DaO_2). During exercise, an appropriate redistribution of blood flow, increases in the number of perfused capillaries in actively respiring tissues and shifts to the right in the oxyhaemoglobin dissociation curve increase the extraction ratio to some theoretical upper limit e.g. 50%–70%. Presumably, this upper limit varies between individuals and depends upon such factors as capillarity of tissues, the haemoglobin concentration and the P_{50}.

(iii) *Maldistributions of flow in the critically ill (line 0,4,7 and the more severe abnormality line 0,6).* In these two settings the appropriate metabolic rate is unknown—it may be normal (through *line 1,4,6*) or it may be increased (*line 2,5,7*). In both cases, because the patient's ability to increase his oxygen extraction ratio is impaired (defective autoregulation), OUI is likely to be limited by DaO_2 values much above the physiological critical DaO_2. It is important to note that patients with a more severe maldistribution assessed over a prolonged period of time will demonstrate a flat relation between DaO_2 and OUI. This has been construed as the normal relation but is not so because metabolic rate (hence OUI) is not constant.

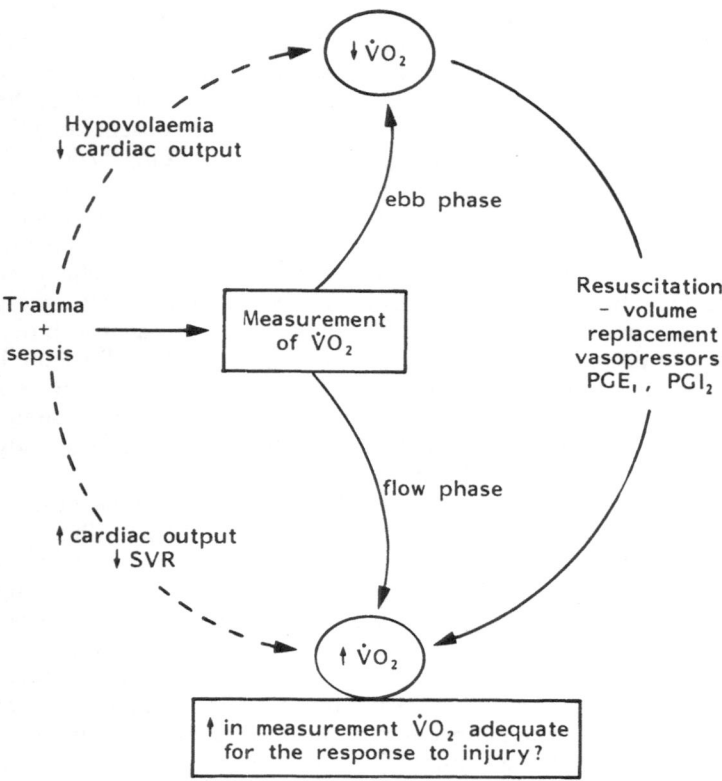

Fig. 7.8. The measurement of oxygen consumption ($\dot{V}O_2$) is central in the assessment of the response to sepsis and trauma. A low $\dot{V}O_2$ usually occurs immediately following injury reflecting hypovolaemia and a reduction in cardiac output. A primary "toxic" reduction of $\dot{V}O_2$ probably does not occur other than in the agonal phases of irreversible shock. Resuscitation (volume replacement, vasoactive compounds such as inotropes, vasopressors and vasodilators) is aimed at improving oxygen delivery to tissues so that any increase in metabolic rate is not limited by the supply of substrate. In the assessment of a critically ill patient the important question is always: "Is the measured oxygen consumption adequate for the response to injury or is it limited by the oxygen supply?"

have demonstrated that the low vascular resistance/high cardiac output state associated with acute liver failure does indeed result in severe tissue hypoxia, despite substantial increases in cardiac output and arterial/mixed venous oxygen tensions (Bihari et al. 1986a, b). This tissue hypoxia appears to be more severe in patients who subsequently die with irreversible multiple organ failure. Acute liver failure, like septic shock, may also be associated with a lactic acidosis (Bihari et al. 1985a) and shifts to the right in the oxyhaemoglobin dissociation curve (Bihari et al. 1985b) consistent with a marked impairment of tissue oxygenation.

Our hypothesis that a disturbance of microcirculatory blood flow—rather than any primary reduction in systemic or hepatic oxidative capacity (Nespoli et al. 1981; Siegel et al. 1981; Loda et al. 1984)—underlies the development of tissue hypoxia in these patients is supported by data obtained from an analysis of the effects of further vasodilatation using prostacyclin (PGI_2). We have studied the effects of increasing

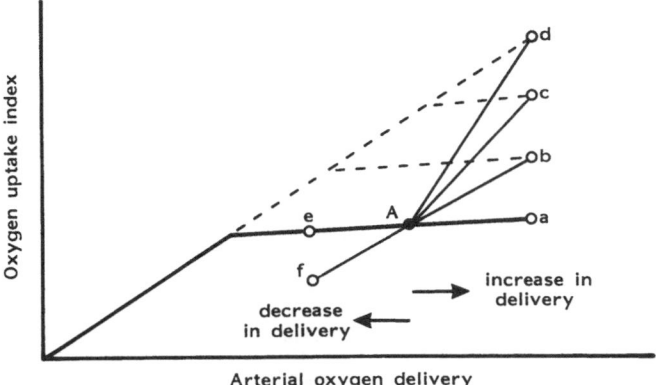

Fig. 7.9. Oxygen flux test: the direct assessment of tissue oxygenation. The best way to assess the adequacy of tissue oxygenation (given apparently adequate values for the blood pressure, urine output, cardiac output, haemoglobin concentration (and its P_{50}), arterial and mixed venous oxygen tensions and saturations, pH, bicarbonate and lactate concentrations) is to increase the delivery of oxygen (DaO_2) to tissues (over a short period of time during which the metabolic rate is constant) and observe the effect on oxygen uptake (OUI).

Baseline measurements of DaO_2 and OUI are obtained (and both may be in the normal physiological range or increased); DaO_2 is then increased (in our studies by the infusion of a vasodilatory prostaglandin, PGI_2). An increase in OUI (given that PGI_2 is not a primary metabolic stimulant) suggests an underlying mitochondrial or extramitochondrial debt, the severity of which may be assessed by the percentage increase in OUI. Hypothetical patients (A–a) have no tissue oxygen debt, (A–b) a measurable debt, (A–c) a substantial debt and (A–d) a lethal debt.

Similarly, a decrease in DaO_2 can be obtained by the application of PEEP. Patients (A–e) (who demonstrate an increase in oxygen extraction ratio) have intact mechanisms of autoregulation undisturbed by PEEP, whereas patients (A–f) may have defective autoregulation so that the lower DaO_2 cannot be appropriately redistributed and is accompanied by a reduction in OUI. This may be the result of the disease process (sepsis, trauma) or the application of PEEP.

oxygen flux by the infusion of PGI_2 at 5 ng kg^{-1} min $^{-1}$ for 30 minutes in 11 patients with acute liver failure and in grade III or IV encephalopathy. We observed that increases in DaO_2 as a result of a further fall in systemic vascular resistance were associated with significant increases in oxygen consumption (Table 7.3, Fig. 7.10). In another 8 patients studied the same dose of PGI_2 over a period of 1 hour produced not only an increase in OUI (medians before and after 134 and 168 ml min $^{-1}$ m^{-2} respectively) but also a fall in the mixed venous lactate concentration (from a median of 3.8 to 3.2 mmol l^{-1}).

Similarly, we have now studied 27 patients with sepsis and dysfunction of two or more organ systems in our general intensive therapy unit and have demonstrated that in both low and high cardiac output states the development of irreversible multiple organ failure and subsequent death may be predicted early on in the course of a patient's illness by the response to an increase in DaO_2 obtained by a 30 minute infusion of PGI_2 (Bihari et al. 1986d). We have found that patients who subsequently die have a significantly greater percentage increase in OUI during the PGI_2 infusion compared with survivors. Furthermore, PGI_2 produced an increase in the oxygen extraction ratio in 10 of the 13 patients who died in contrast with the reductions that were observed in all normal controls and surviving patients.

Table 7.3. The effects of PGI_2 on systemic haemodynamics and oxygen transport in 11 patients with fulminant hepatic failure (medians with ranges)

	Before PGI_2 infusion ($n = 14$)	During PGI_2 infusion ($n = 14$)
Mean arterial pressure (mmHg)	87 (55–116)	90 (41–110)
Cardiac index, CI ($l\,min^{-1}\,m^{-2}$)	4.5 (2.2–7.1)	5.4 (2.5–9.3)[a]
Systemic vascular resistance ($dyne\cdot s\,cm^{-5}\,m^{-2}$)	1739 (652–3103)	1257 (533–2995)[a]
Arterial O_2 tension (mmHg)	92 (74–232)	83 (53–228)[b]
Arterial O_2 delivery, DaO_2 ($ml\,O_2\,min^{-1}\,m^{-2}$)	657 (373–1114)	786 (498–1360)[a]
Oxygen uptake index, OUI ($ml\,O_2\,min^{-1}\,m^{-2}$)	134 (94–256)	164 (124–247)[a]

[a] $p<0.01$.
[b] $p<0.02$.

One interpretation of these observations is that in the more critically ill patients, PGI_2 not only vasodilated regions of the microcirculation that were inappropriately vasoconstricted, thus revealing a "covert" oxygen debt, but also improved the distribution of the oxygen supply so that a greater percentage of delivered oxygen was taken up. These data provide strong support for unrecognized tissue hypoxia arising

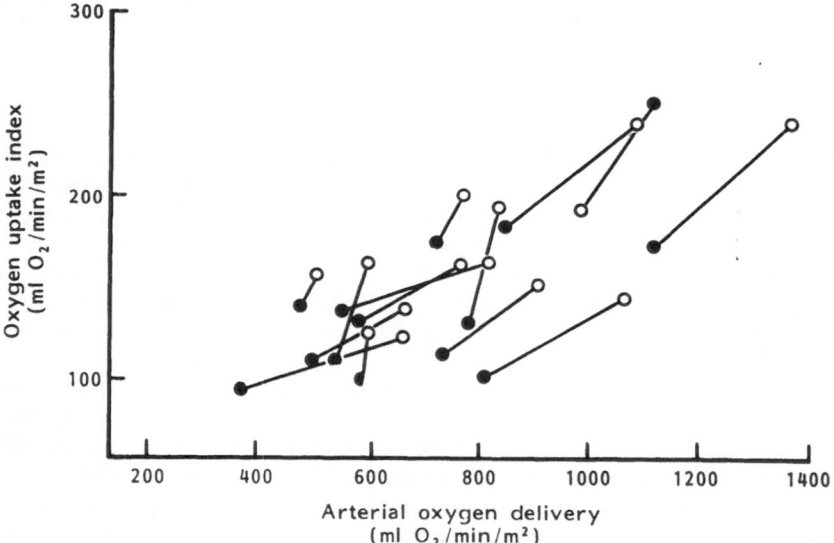

Fig. 7.10. The effects of PGI_2 (5 ng kg^{-1} min^{-1}) on arterial oxygen delivery (DaO_2) and oxygen uptake index (OUI) in 11 patients with acute liver failure and in grade III or IV encephalopathy, studied on 14 occasions during the course of their illness. ●, before PGI_2; ○, during PGI_2 infusion. One patient had a substantial fall in cardiac filling pressures and a fall in DaO_2 accompanied by a reduction in OUI.

from defective microcirculatory autoregulation as a central mechanism mediating the widespread and irreversible organ damage that occurs in some cases of severe sepsis and liver failure. Nevertheless, it remains to be seen whether a more prolonged PGI_2 infusion will prevent the development of multiple organ failure; such a study is now in progress.

Problems in the Interpretation of the Oxygen Flux Test

We have suggested that only with some form of "oxygen flux test" rather than any other measurement of systemic oxygenation (arterial/mixed venous oxygen tension or cardiac output) can the presence of tissue hypoxia be identified. Moreover, it may be that with the early identification and treatment of such a tissue oxygen debt the incidence of multiple organ failure in the critically ill might be reduced. Yet there remain two potential problems with this approach: firstly, the means used to increase DaO_2 may be of critical importance in determining the changes in OUI observed. Secondly, there is the possibility that any extra oxygen taken up by a critically ill patient might not be usefully employed—that is, one may be observing either the uncoupling of oxidative phosphorylation or an increased generation of oxygen free radicals which might contribute further to tissue damage (Cain 1984). This last suggestion can probably be discounted for the quantities of oxygen required by white cells for free radical formation are extremely small (less than 1% of OUI) and could not readily be detected in the whole body preparation (Anthony Segal, pers. comm.).

Oxygen consumption that is independent of ATP production, i.e. that occurs as a result of the uncoupling of oxidative phosphorylation, is more difficult to discount, but most agents which produce uncoupling (e.g. dinitrophenol) do in fact produce a fall in oxygen consumption. Moreover, in some of our patients with acute liver failure we documented an early "washout" effect, with the blood lactate level initially rising and then falling concomitant with the increase in OUI associated with the PGI_2 infusion. Other investigators have emphasized that "delivery-dependent oxygen consumption" is more likely to occur in the presence of a lactic acidosis (Haupt et al. 1985; Rashkin et al. 1985) which may be corrected in some cases by increasing DaO_2 by adequate volume expansion alone (Kaufman et al. 1984). These data support the contention that the extra oxygen taken up is being used usefully in those patients with lactic acidaemia but do not detract from the earlier suggestion that in some patients with a normal pH and blood lactate concentration the correction of an extra-mitochondrial oxygen debt (independent of the presence or absence of lactate) might be just as important in maintaining organ function.

Catecholamines and their synthetic sympathomimetic derivatives (especially dobutamine) have been used to increase DaO_2 in various groups of critically ill patients but results from these studies are more difficult to interpret because of the primary effect of adrenoreceptor stimulation on intrinsic metabolic rate (McDonald et al. 1985). It may be that adrenaline, dopamine and dobutamine produce a primary increase in OUI (especially in myocardial oxygen consumption) that must be met by an appropriate increase in DaO_2. In these studies a fall in the oxygen extraction ratio might be better evidence of an improvement in the supply/demand relationship. Similarly, increasing arterial oxygen tension (by manipulations of the inspired oxygen concentration, the I:E ratio or PEEP/CPAP) or the haemoglobin concentration may have variable effects on tissue oxygenation because oxygen per se is an arteriolar vasoconstrictor (Duling and Pittman 1975) and changes in haemoglobin influence whole blood viscosity, which is an important determinant of the distribution of micro-circulatory blood flow (Goslinga 1984)

It would seem that vasodilatation is probably the most appropriate means of carry-ing out an oxygen flux test but it is not clear whether all vasodilators act in the same way or whether some compounds (e.g. nitroprusside), despite increasing the cardiac output and oxygen despatch, exaggerate any maldistribution of flow. Certainly, in animal studies there is evidence to suggest that alpha blockade impairs autoregula-tion so that oxygen consumption becomes dependent upon DaO_2 at a significantly higher level of DaO_2 (Cain 1978). Nevertheless, Shoemaker has emphasized from his controlled trial of survival in the postoperative critically ill that it is important to maintain DaO_2 and OUI at supranormal levels (which presumably was achieved in this study by using volume loading and inotropic support) in order to reduce the inci-dence of multiple organ failure and overall mortality (Shoemaker et al. 1982).

The Therapeutic Role of Vasodilator Prostaglandins in Sepsis

Our observations of the effect of PGI_2 on tissue oxygenation are supported by a large body of evidence in the literature concerning the role of PGI_2 and TXA_2 in the pathogenesis of organ dysfunction associated with septic shock (Hechtman et al. 1983; Ball et al. 1986). Although it is likely, as stated earlier, that a number of differ-ent mediator systems are responsible for the tissue damage that occurs following an episode of septic or traumatic shock, several investigators have demonstrated elevated plasma TXB_2 and 6-keto-$PGF_{1\alpha}$ levels in various animal models of endotoxaemia and trauma (Cook et al. 1980; Webb et al. 1981; Carmona et al. 1984). Furthermore, Halushka's group have reported markedly elevated levels of both these metabolites in patients dying with septic shock compared with survivors and normal controls (Reines et al. 1982; Halushka et al. 1985). Others have observed significantly higher concentrations of TXB_2 in patients developing acute respiratory failure associated with sepsis (Lamy et al. 1985).

Endotoxin-induced lung injury in sheep with a chronic lung lymph fistula is one particular animal model that has been extensively studied and both the increases in pulmonary capillary permeability and pulmonary artery pressure that occur in this model are closely related in time to the appearance of TXB_2 and 6-keto-$PGF_{1\alpha}$ in lung lymph (Demling et al. 1981a; Brigham 1985). More recently, interest has centred on the role of the various metabolites of arachidonic acid produced by the 5,12- and 15-lipoxygenase enzyme present in polymorphonuclear leucocytes and macrophages (Gee et al. 1985; Malik et al. 1985; Anonymous 1986a). Activation and sequestration of both white cells and platelets within the pulmonary microcirculation seem to be important mechanisms in the generation of this acute lung injury (Smith et al. 1981). Treatment of animals with a PGI_2 infusion not only reduces lung lymph flow and lysosomal enzyme concentrations but also improves lung function and survi-val (Demling et al. 1981b; Smith et al. 1982). Similarly, PGI_2 may reverse lethal endotoxaemia in dogs (Krausz et al. 1981) and prevent the lung injury induced in these animals by oleic acid (Slotman et al. 1982). The administration of a selective

TXA_2 synthetase inhibitor has a protective effect in animal studies of endotoxic shock (Wise et al. 1980; Halushka et al. 1983) and has also been shown to maintain the autoregulation of blood flow in response to adenosine in an isolated canine hindlimb preparation treated with microemboli (Shah et al. 1981).

Other than the open studies on sepsis and acute liver failure previously described, there are, at present, no controlled data available concerning the effect of a PGI_2 infusion on survival in the critically ill human. However, the results of a North American controlled clinical trial of the similar compound, PGE_1, in patients with acute respiratory failure associated with trauma and sepsis have recently been presented (Holcroft et al. 1986). A 7 day continuous infusion of PGE_1 (30 ng kg^{-1} min^{-1}) was associated with a significantly improved 30 day survival: 15 of 21 (71%) treated patients survived compared with only 7 of the 20 (35%) controls. Most significantly, of those PGE_1-treated patients who had acute respiratory failure but no other organ dysfunction, none developed multiple organ failure and all survived. Both Shoemaker (Shoemaker and Appel 1986) and a Japanese group (Tokioka et al. 1985) have demonstrated that PGE_1 (as does PGI_2) not only reduces pulmonary vascular resistance but also increases DaO_2 and oxygen consumption in this group of patients, and on the basis of these results a multi-centre prospective randomized trial of PGE_1 in acute respiratory failure associated with sepsis or trauma is now under way in the United States and Europe.

Interestingly, there has been no direct comparison of PGE_1 with PGI_2 in patients with acute respiratory failure, although two animal studies have suggested that PGI_2 is more effective in the prevention of acute lung injury (Slotman et al. 1982; Smith et al. 1982). Both prostaglandins have a number of actions that might be beneficial in the septic state (Table 7.4) and in particular, PGI_2 has been shown to reduce monocyte and macrophage activation, reducing the release of various monokines (Needleman and Tripp 1986). The anti-platelet effect of PGE_1 only occurs at very high doses and as PGE_1 is metabolized in the lungs, its systemic effects are greatly reduced. However, although hypotension is less likely with this compound, it may be that systemic effects on other organs are particularly helpful. We have emphasized the improvement in tissue oxygenation observed with PGI_2, but one other important

Table 7.4. Possible beneficial effects of PGE_1 and PGI_2 in the prevention and management of multiple organ failure associated with sepsis and trauma

---

1. Improvement in microcirculatory blood flow with the prevention of tissue hypoxia:
    Microcirculatory vasodilatation
    Inhibition of platelet activation/aggregate formation with thromboxane A_2 release
    Fibrinolysis (reported with PGI_2 only)

2. Inhibition of polymorphonuclear leucocyte activation (PGE_1) and adhesion to vascular endothelium (PGI_2)

3. Inhibition of monocyte activation and the release of monokines (PGE_2 and PGI_2)

4. "Cytoprotection" (reported with PGE_1, PGE_2, 15,15-dimethyl-PGE_2 and PGI_2)

5. Pulmonary vasodilation with a reduction in pulmonary artery pressure and calculated resistance

6. Increased delivery of antimicrobial agents to septic foci

---

aspect of the function of these prostaglandins is the well-described phenomenon of "cytoprotection" (Robert 1979). This action, which may (Leung et al. 1985) or may not (Gaskill et al. 1982) be independent of changes in blood flow, has been described in many different in vitro organ systems (Araki and Lefer 1980; Ruwart et al. 1981; Stachura et al. 1981; Utsunomiya et al. 1982; Sikujara et al. 1983; Noda et al. 1986) and may play an important part in the prevention of multiple organ failure in treated patients.

Conclusion

The prevention and treatment of multiple organ failure associated with trauma and sepsis has become the major challenge facing the intensive care physician. Mortality remains unacceptably high despite the introduction of sophisticated mechanical and pharmacological support systems. We suggest that one possible mechanism of organ failure in these critically ill patients is a disturbance of microcirculatory autoregulation so that tissue oxygen consumption becomes dependent on and, in some cases, limited by blood flow. This may lead to an oxygen debt which is frequently unrecognized and therefore untreated. Prostacyclin (PGI_2) may, by a variety of different actions, improve microcirculatory blood flow and hence tissue oxygenation. However, it remains to be seen whether such manipulations of the microcirculation improve overall survival.

I would like to thank Drs. Alexander Gimson, Roger Williams and Jack Tinker for their enthusiastic support for many of these ideas. I wish to acknowledge the friendship and assistance of the many different residents and fellows with whom it has been a great pleasure to work in the Liver Failure Unit, King's College Hospital, and the Intensive Therapy Unit, The Middlesex Hospital. Finally, and most importantly, I am most grateful to the nursing staff of both units who by their goodwill have made many of these studies possible. Undoubtedly, their expertise and dedication have been the most important factors in our patients' survival.

References

Adams R, Dieleman L, Cain S (1982) A critical value for oxygen transport in the rat. J Appl Physiol 53:660–664
Andreadis N, Petty T (1985) Adult respiratory distress syndrome: problems and progress. Am Rev Respir Dis 132:1344–1346
Anonymous (1984) Ketone body ratio: an index of multiple organ failure. Lancet I:25–26
Anonymous (1986a) Adult respiratory distress syndrome. Lancet I:301–303
Anonymous (1986b): ARDS: a clinical view. Lancet II:439
Araki H, Lefer A (1980) Cytoprotective actions of prostacyclin during hypoxia in the isolated perfused cat liver. Am J Physiol 238:H176–H181
Askanazi J, Forse R, Weissman C, Hyman A, Kinney J (1986) Ventilatory effects of the stress hormones in normal man. Crit Care Med 14:602–605
Ball H, Cook W, Wise W, Halushka P (1986) Role of thromboxane, prostaglandins and leucotrienes in endotoxic and septic shock. Intensive Care Med 12:116–126

Baracos V, Rodemann H, Dinarello C, Goldberg A (1983) Stimulation of muscle protein degradation and PGE$_2$ release by leucocytic pyrogen (interleukin 1). N Engl J Med 308:553–558

Barcroft J (1920) Physiological effects of insufficient oxygen supply. Nature (Lond) 106:125–129

Bartlett R, Dechert R, Mault J, Ferguson A, Kaiser A, Erlandson E (1982) Measurement of metabolism in multiple organ failure. Surgery 92:771–779

Baue A (1975) Multiple, progressive or sequential systems failure: a syndrome of the 1970s. Arch Surg 110:779–781

Bell R, Coalson J, Smith J, Johanson W (1983) Multiple organ system failure and infection in adult respiratory distress syndrome. Ann Intern Med 99:293–298

Bernton E, Long J, Holaday J (1985) Opioids and neuropeptides: mechanisms in circulatory shock. Fed Proc 44:290–295

Berry M, Scheuer J (1967) Splanchnic lactic acid metabolism in hyperventilation, metabolic alkalosis and shock. Metabolism 16:537–547

Bertrand Y (1985) Oxygen free radicals and lipid peroxidation in the adult respiratory distress syndrome. Intensive Care Med 11:56–60

Bihari D (1985) Acute liver failure. Clin Anaesthesiol 3:973–997

Bihari D, Gimson A (1983) Oxygen delivery, mixed venous oxygenation and haemodynamics in chronic obstructive pulmonary disease. N Engl J Med 309:1251 (letter)

Bihari D, Tinker J (1983) The management of shock. In: Tinker J, Rapin M (eds) Care of the critically ill patient. Springer-Verlag, Berlin Heidelberg New York Tokyo, pp 189–222

Bihari D, Gimson A, Lindridge J, Williams R (1985a) Lactic acidosis in fulminant hepatic failure: some aspects of pathogenesis and prognosis. J Hepatol 1:405–416

Bihari D, Gimson A, Waterson M, Williams R (1985b) Tissue hypoxia in fulminant hepatic failure. Crit Care Med 13:1034–1039

Bihari D, Gimson A, Williams R (1986a) Disturbances in cardiovascular and pulmonary function in fulminant hepatic failure. In: Williams R (ed) Liver failure. Clinics in Critical Care Medicine. Churchill Livingstone, Edinburgh, pp 47–71

Bihari D, Gimson A, Williams R (1986b) Cardiovascular, pulmonary and renal complications of fulminant hepatic failure. Semin Liver Dis 6:119–128

Bihari D, Smithies M, Gimson A (1986c) The pathogenesis of multiple organ failure associated with septicaemic shock. Curr Clin Concepts 3:49–60

Bihari D, Smithies M, Tinker J (1986d) Unrecognised tissue hypoxia in the critically ill: its relation to survival. Crit Care Med 14:349

Bihari D, Smithies M, Pozniak A, Gimson A (1987) A comparison of direct and indirect measurements of oxygen delivery and consumption in human volunteers and critically ill patients: the effects of prostacyclin. Scand J Clin Lab Invest 47 [Suppl 188]:37–45

Bland J, Altman D (1986) Statistical methods for assessing agreement between two methods of clinical measurement. Lancet I:307–310

Bond R, Johnson G (1985) Vascular adrenergic interactions during hemorrhagic shock. Fed Proc 44:281–289

Border J, Gallo E, Schenk W (1966) Systemic arteriovenous shunts in patients under severe stress: a common cause of high output cardiac failure. Surgery 60:225–231

Borzotta A, Polk H (1983) Multiple organ system failure. Surg Clin North Am 63:315–336

Brent B, Matthay R, Mahler D, Berger H, Zaret B, Lister G (1984) Relationship between oxygen uptake and oxygen transport in stable patients with chronic obstructive pulmonary disease. Am Rev Respir Dis 129:682–686

Brigham K (1985) Metabolites of arachidonic acid in experimental lung vascular injury. Fed Proc 44:43–45

Brigham K, Meyrick B (1986) Endotoxin and lung injury. Am Rev Respir Dis 133:913–927

Burns H, Cowan B, Ledingham I (1982) Metabolic acidosis in the critically ill. In: Porter R, Lawrenson G (eds) Metabolic acidosis. Ciba Foundation Symposium 87. Pitman, London, pp 293–306

Cain S (1977) Oxygen delivery and uptake in dogs during anaemic and hypoxic hypoxia. J Appl Physiol 42:228–234

Cain S (1978) Effects of time and vasoconstrictor tone on oxygen extraction during hypoxic hypoxia. J Appl Physiol 45:219–224

Cain S (1984) Supply dependency of oxygen uptake in ARDS: myth or reality? Am J Med Sci 288:119–124

Cameron J (1986) Acute renal failure in the Intensive Care Unit today. Intensive Care Med 12:64–70

Carlsson M, Nordenstrom J, Hedendtierna G (1984) Clinical implications of continuous measurement of energy expenditure in mechanically ventilated patients. Clin Nutrition 3:103

Carmona R, Tsao T, Trunkey D (1984) The role of prostacyclin and thromboxane in sepsis and septic shock. Arch Surg 119:189–198

Carroll G, Snyder J (1980) Peripheral anatomic systemic venous shunting in high cardiac output septic shock. Crit Care Med 8:230

Cerra F, Siegel J, Border J, Wiles J, McMenamy R (1979) The hepatic failure of sepsis. Surgery 86:409–422

Chappell T, Rubin L, Markham R, Firth B (1983) Independence of oxygen consumption and systemic oxygen transport in patients with either stable pulmonary hypertension or refractory left ventricular failure. Am Rev Respir Dis 128:30–33

Chaudry I, Clemens M, Baue A (1986) Cellular and subcellular function of the liver and other vital organs in sepsis and septic shock. In: Sibbald W, Sprung C (eds) Perspectives on sepsis and septic shock. New horizons vol 1. Society of Critical Care Medicine, California, pp 61–76

Clowes G, George B, Villee C, Saravis C (1983) Muscle proteolysis induced by a circulating peptide in patients with sepsis and trauma. N Engl J Med 308:545–552

Coalson J (1986) Pathology of sepsis, septic shock and multiple organ failure. In: Sibbald W, Sprung C (eds) Perspectives on sepsis and septic shock. New horizons vol 1. Society of Critical Care Medicine, California, pp 27–60

Cohen R, Woods H (1976) Clinical and biochemical aspects of lactic acidosis. Blackwell Scientific Publications, Oxford

Cohen R, Woods H (1983) Lactic acidosis revisited. Diabetes 32:181–191

Cohn J, Greenspan M, Goldstein C, Gudwin A, Siegel J, Del Guercio L (1968) Arteriovenous shunting in high cardiac ouput shock syndromes. Surg Gynaecol Obstet 127:282–288

Cook J, Wise W, Halushka P (1980) Elevated thromboxane levels in the rat during endotoxic shock. J Clin Invest 65:227–230

Cowan B, Burns H, Boyle P, Ledingham I (1984) The relative prognostic value of lactate and haemodynamic measurements in early shock. Anaesthesia 39:750–755

Cryan L, Ledingham I (1986) The significance of blood lactate in Intensive Care. Intensive and Critical Care Digest 5:15–17

Danek S, Lynch J, Weg J, Dantzker D (1980) The dependence of oxygen uptake on oxygen delivery in adult respiratory distress syndrome. Am Rev Respir Dis 122:387–395

Daniel A, Cohen J, Lichtman M, Murphy M, Schreiner B, Shah P (1978) The relationships among arterial oxygen flow rate, oxygen binding by hemoglobin, and oxygen utilisation in chronic cardiac decompensation. J Lab Clin Med 91:635–649

Davila F (1984) Correlations of independently derived components of calculated oxygen consumption and oxygen delivery. Am Rev Respir Dis 129:A96 (abstr)

Demling R, Smith M, Gunther R, Flynn J, Gee M (1981a) Pulmonary injury and prostaglandin production during endotoxinaemia in conscious sheep. Am J Physiol 240:H348–H353

Demling R, Smith M, Gunther R, Gee M, Flynn J (1981) The effect of a prostacyclin infusion on endotoxin-induced lung injury. Surgery 89:257–263

Denison D (1981) The distribution and use of oxygen in tissues. In: Scadding J, Cumming G, Thurlbeck W (eds) The scientific foundations of respiratory medicine. Heinemann, London, pp 221–237

Duff J, Groves A, McLean A, LaPointe R, MacLean L (1969) Defective oxygen consumption in septic shock. Surg Gynaecol Obstet 128:1051–1060

Duling B, Pittman R (1975) Oxygen tension: dependent or independent variable in local control of blood flow? Fed Proc 34:2012–2019

Faist E, Baue A, Dittmer H et al. (1983) Multiple organ failure in polytrauma patients. J Trauma 19:305–318

Filkins J (1985) Monokines and the metabolic pathophysiology of septic shock. Fed Proc 44:300–304

Finley R, Duff J, Holliday R, Jones D, Marchuk J (1975) Capillary muscle blood flow in human sepsis. Surgery 78:87–94

Fry DE, Pearlstein L, Fulton RL et al. (1980) Multiple system organ failure. The role of uncontrolled infection. Arch Surg 115:136–140

Gaskill H, Sinnek K, Levine B (1982) Prostacyclin mediated gastric cytoprotection is dependent upon mucosal blood flow. Surgery 92:220–225

Gee M, Perkowski S, Tahamont M, Flynn J (1985) Arachidonate cyclooxygenase metabolites as mediators of complement-initiated lung injury. Fed Proc 44:46–52

George R, Tinker J (1983) The pathogenesis of shock. In: Tinker J, Rapin M (eds) Care of the critically ill patient. Springer-Verlag, Berlin Heidelberg New York Tokyo, pp 168–188

Gimson A (1987) Hepatic dysfunction during bacterial sepsis. Intensive Care Med (in press)

Gimson A, Bihari D, Wilson C, Williams R (1984) Delivery dependent oxygen consumption in acute liver failure. Clin Sci 66:12P

Goslinga H (1984) Blood viscosity and shock. Springer-Verlag, Berlin Heidelberg New York Tokyo pp 101–102

Grainger H, Goodman A, Cook B (1975) Metabolic models of microcirculatory regulation. Fed Proc 34:2025–2030

Grum C, Simon R, Dantzker D, Fox I (1985) Evidence for adenosine triphosphate degradation in critically ill patients. Chest 88:763–767

Halushka P, Cook J, Wise W (1983) Beneficial effects of UK 37248, a thromboxane synthetase inhibitor, in experimental endotoxic shock in the rat. Br J Clin Pharmacol 15:1335–1338

Halushka P, Reines D, Barrow S, Blair I, Dollery C, Rambo W, Cook J, Wise W (1985) Elevated plasma 6-keto-prostaglandin F in patients with septic shock. Crit Care Med 13:451–453

Hasselgren P-O, Fischer J (1986) Septic encephalopathy: etiology and management. Intensive Care Med 12:13–16

Haupt M, Gilbert E, Carlson R (1985) Fluid loading increases oxygen consumption in septic patients with lactic acidosis. Am Rev Respir Dis 131:912–916

Hechtman H, Huval W, Mathieson M, Stemp L, Valeri C, Shepro D (1983) Prostaglandin and thromboxane mediation of cardiopulmonary failure. Surg Clin North Am 63:263–283

Holcroft J, Vassar M, Weber C (1986) Prostaglindin E_1 and survival in patients with adult respiratory distress syndrome. Ann Surg 203:371–378

Houtchens B, Westenskow D (1984) Oxygen consumption in septic shock: collective review. Circ Shock 13:361–384

Jones G, Hurley J (1984) The effect of prostacyclin on the adhesion of leucocytes to injured vascular endothelium. J Pathol 142:51–59

Kariman K, Burns S (1985) Regulation of tissue oxygen extraction is disturbed in adult respiratory distress syndrome. Am Rev Respir Dis 132:109–114

Kaufman B, Rackow E, Falk J (1984) The relationship between oxygen delivery and consumption during fluid resuscitation of hypovolaemic and septic shock. Chest 85:336–340

Knaus W, Draper E, Wagner D, Zimmerman J (1985) Prognosis in acute organ-system failure. Ann Surg 202:685–693

Knight A, Bihari D, Tinker J (1985) Stress ulceration in the critically ill patient. Br J Hosp Med 33:216–219

Krausz M, Utsunomiya T, Feuerstein G, Wolfe J, Shepro D, Hechtman H (1981) Prostacyclin reversal of lethal endotoxaemia in dogs. J Clin Invest 67:1118–1125

Lamy M, Deby-Dupont G, Braun M, Faymonville M, Princemail J, Deby C, van Erck J, Damas P (1985) Thromboxane, prostacyclin, PGE_2 and the human adult respiratory distress syndrome. In: Proceedings of the 4th world congress on intensive and critical care medicine. Jerusalem, p 39 (abstr 112)

Ledingham I, Cown B, Burns H (1982) Prognosis in severe shock (editorial). Br Med J 284:443

Lefer A (1985) Eicosanoids as mediators of ischaemia and shock. Fed Proc 44:275–280

Leung F, Robert A, Guth P (1985) Gastric mucosal blood flow in rats after the administration of 16,16-dimethyl PGE_2 at a cytoprotective dose. Gastroenterology 88:1948–1953

Loda M, Clowes G, Nespoli A, Bigatello L, Birkett D, Menzoian J (1984) Encephalopathy, oxygen consumption, visceral amino-acid clearance and mortality in cirrhotic surgical patients. Am J Surg 147:542–550

Long G, Sznajder J, Nelson D, Fan P, Schumacher P, Wood L (1984) The independent influence of acute lung injury and PEEP on the relationship between oxygen delivery and consumption in dogs. Am Rev Respir Dis 129:A97

Malik A (1983) Pulmonary microembolism. Physiol Rev 63:1115–1195

Malik A, Perlman M, Cooper J, Nooman T, Bizois R (1985) Pulmonary microvascular effects of arachidonic acid metabolites and their role in lung vascular injury. Fed Proc 44:36–42

Mann S, Westenskow D, Houtchens B (1985) Measured and predicted caloric expenditure in the acutely ill. Crit Care Med 13:173

McDonald I, Bennett T, Fellows I (1985) Catecholamines and the control of metabolism in man. Clin Sci 68:613–619

Mela L (1983) Mitochondrial function in shock, ischemia and hypoxia. In: Cowley R, Trump B (eds) Pathophysiology of shock, anoxia and ischemia. Williams and Wilkins, Baltimore

Mela L, Bacalzo L, Miller L (1971) Defective oxidative metabolism of rat liver mitochondria in haemorrhage and endotoxin shock. Am J Physiol 220:571–577

Miller M (1982) Tissue oxygenation in clinical medicine: an historical review. Anaesth Analg (Cleveland) 61:527–535

Mohsenifar Z, Goldbach P, Tashkin D, Campisi D (1983) Relationship between oxygen delivery and consumption in adult respiratory distress syndrome. Chest 84:267–271

Morrison D, Ulevitch R (1978) The effects of bacterial endotoxins on host mediation systems. Am J Pathol 93:527–617

Needleman P, Tripp C (1986) The regulation and function of macrophage eicosanoid metabolism in tissue injury. In: Abstracts of the 6th international conference on prostaglandins and related compounds, Florence, p 484

Nelson L, Houtchens B, Westonskow D (1982) Oxygen consumption and optimum PEEP in acute respiratory failure. Crit Care Med 10:857–862

Nespoli A, Berilacqua G, Staudacher C, Rossi N, Salerno F, Castelli M (1981) Pathogenesis of hepatic encephalopathy and the hyperdynamic syndrome in cirrhosis: the role of false neurotransmitters. Arch Surg 116:1129–1138

Noda Y, Hughes R, Williams R (1986) Effects of prostacyclin and a prostaglandin analogue BW 245C on galactosamine induced hepatic necrosis. J Hepatol 2:53–64

Nolan J (1981) Endotoxin, reticuloendothelial cell function and liver injury. Hepatology 1:458–465

Nowak J, Weunmalm A (1978) Influence of indomethacin and PGE_1 on total and regional blood flow in man. Acta Physiol Scand 102:484–491

Old L (1985) Tumour necrosis factor. Science 230:630–632

Ozawa K, Aoyama H, Shimahara Y, Nakatani T, Tanaka J, Yamamoto M, Kamiyaoma Y, Tobe T (1983) Metabolic abnormalities associated with postoperative organ failure. Arch Surg 118:1245–1251

Park R, Arieff A (1983) Lactic acidosis: current concepts. Clin Endocrinol Metab 12:339–358

Parker M, Shelhamer J, Bacharach S et al. (1984a) Profound but reversible myocardial depression in patients with septic shock. Ann Intern Med 100:483–490

Parker M, Shelhamer J, Natanson C, Masur H, Parrillo J (1984b) Serial haemodynamic patterns in survivors and nonsurvivors of septic shock in humans. Crit Care Med 12:311 (abstr)

Pepe P, Culver B (1982) Independently measured oxygen consumption and oxygen delivery in acute lung injury. Chest 82:248–249

Petty T (1985) Indicators of risk, course and prognosis in adult respiratory distress syndrome (editorial). Am Rev Respir Dis 132:471

Powers S, Mannal R, Neclerio M et al. (1973) Physiological consequences of PEEP. Ann Surg 178:265–272

Rashkin M, Bosken C, Baughman R (1985) Oxygen delivery in critically ill patients: relationship to blood lactate and survival. Chest 87:580–584

Reines H, Halushka P, Cook J, Wise W, Rambo W (1982) Plasma thromboxane concentrations are raised in patients dying with septic shock. Lancet II:174–175

Rhodes R, Newell J, Shah D, Scovill W, Tauber J, Dutton R, Powers S (1978) Increased oxygen consumption accompanying increased oxygen delivery with hypertonic mannitol in adult respiratory distress syndrome. Surgery 84:490–497

Robert A (1979) Cytoprotection by prostaglandins. Gastroenterology 77:761–777

Robin E (1980) Of men and mitochondria: coping with hypoxic dysoxia. Am Rev Respir Dis 122:517–531

Rubin S, Siemienczuk D, Nathan M, Prause J, Swan H (1982) Accuracy of cardiac output, oxygen uptake and arterio-venous oxygen difference at rest, during exercise and after vasodilator therapy in patients with severe, chronic heart failure. Am J Cardiol 50:973–978

Ruwart M, Rush B, Friedle N, Piper R, Kolaja G (1981) Protective effects of 16,16-dimethyl PGE_2 on the liver and kidney. Prostaglandins 21 [Suppl]:97–102

Saba T (1986) Organ failure with sepsis after trauma or burn: support of the reticuloendothelial host defense system. In: Sibbald W, Sprung C (eds) Perspectives on sepsis and septic shock. New horizons vol 1. Society of Critical Care Medicine, California, pp 77–96

Saldeen T (1976) The microembolism syndrome. Microvasc Res 11:227–259

Schumacker P, Cain S (1987) The concept of a critical oxygen delivery. Intensive Care Med (in press)

Schumacker P, Wood L (1984) Limitations of aerobic metabolism in critical illness. Chest 85:453–454

Schumer W, Das Gupta T, Moss G, Nyhus L (1970) Effects of endotoxaemia on liver cell mitochondria in man. Ann Surg 171:875–882

Scovill W, Saba T, Blumenstock F, Bernard H, Powers S (1978) Opsonic alpha-2 surface binding glycoprotein therapy during sepsis. Ann Surg 188:521–529

Seidenfeld J, Pohl D, Bell R, Harris G, Johanson W (1986) Incidence, site and outcome of infections in patients with adult respiratory distress syndrome. Am Rev Respir Dis 134:12–16

Shah D, Newell J, Saba T (1981) Defects in peripheral oxygen utilisation following trauma and shock. Arch Surg 171:1277–1281

Shibutani K, Komatsu T, Kubal K, Sanchala V, Kumar K, Bizzarri D (1983) Critical level of oxygen delivery in anaesthetised man. Crit Care Med 11:640–643

Shoemaker W (1986) Haemodynamic and oxygen transport patterns in septic shock: physiological mechanisms and therapeutic implications. In: Sibbald W, Sprung C (eds) Perspectives on sepsis and septic shock. New horizons vol 1. Society of Critical Care Medicine, California, pp 203–234

Shoemaker W, Appel P (1986) Effects of PGE$_1$ in the adult respiratory distress syndrome. Surgery 99:275–282

Shoemaker W, Czer L (1979) Evaluation of the biologic importance of various haemodynamic and oxygen transport variables. Crit Care Med 7:424–431

Shoemaker W, Appel P, Kram H et al. (1982) Clinical trial of an algorithm for outcome prediction in acute circulatory failure. Crit Care Med 10:390

Sibbald W, Driedger A (1986) Specific organ function/dysfunction in sepsis and septic shock: cardiovascular. In: Sibbald W, Sprung C (eds) Perspectives on sepsis and septic shock. New horizons vol 1. Society of Critical Care Medicine, California, pp 125–146

Siegel J, Greenspan M, Del Guercio L (1967) Abnormal vascular tone, defective oxygen transport and myocardial failure in human septic shock. Ann Surg 165:504–517

Siegel J, Goldwyn R, Farrell M, Gallin P, Friedman H (1974) Hyperdynamic states and the physiologic determinants of survival. Arch Surg 108:282–292

Siegel J, Giovannini I, Coleman B, Cerra F, Nespoli A (1981) Death after portal decompressive surgery: physiological state, metabolic adequacy and the sequence of development of the physiological determinants of survival. Arch Surg 116:1330–1341

Sikujara O, Monden M, Toyoshima K, Okamura J, Kosaki G (1983) Cytoprotective effects of prostaglandin I$_2$ on ischaemia induced hepatic cell injury. Transplantation 36:238–243

Slotman G, Machiedo G, Casey K, Lyons M (1982) Histological and haemodynamic effects of prostacyclin and prostaglandin E$_1$ following oleic lung infusion. Surgery 92:93–100

Smith M, Gunther R, Gee M, Flynn J, Demling R (1981) Leucocytes, platelets and thromboxane A$_2$ in endotoxin-induced lung injury. Surgery 90:102–107

Smith M, Gunther R, Zaiss C, Demling R (1982) Prostaglandin infusion and endotoxin-induced lung injury. Arch Surg 117:175–180

Stachura J, Tarnawski A, Ivey K, Mach T, Bogdal J, Szczudrawa J, Klimczyk B (1981) Prostaglandin protection of carbon tetrachloride induced liver cell necrosis in the rat. Gastroenterology 81:211–217

Till G, Ward P (1986) Systemic complement activation and acute lung injury. Fed Proc 45:13–18

Tilney NL, Bailey GL, Morgan AP (1973) Sequential system failure after rupture of abdominal aortic aneurysms: an unsolved problem in postoperative care. Ann Surg 178:117–122

Tokioka H, Kobayashi O, Ohta Y, Wakabayashi T, Kosaka F (1985) The acute effects of prostaglandin E$_1$ on the pulmonary circulation and oxygen delivery in patients with the adult respiratory distress syndrome. Intensive Care Med 11:61–64

Utsunomiya T, Krausz M, Kobayashi M, Shepro D, Hechtman H (1982) Myocardial protection with prostacyclin after lethal endotoxaemia. Surgery 92:101–108

Wardle E (1982) Acute renal failure in the 1980s: the importance of septic shock and of endotoxaemia. Nephron 30:193–200

Warshawski F, Sibbald W, Driedger A, Cheung H (1986) Abnormal neutrophil–pulmonary interaction in the adult respiratory distress syndrome. Am Rev Respir Dis 133:797–804

Webb P, Westwick J, Scully M, Zahari J, Kakkar V (1981) Do prostacyclin and thromboxane play a role in endotoxic shock? Br J Surg 68:720–724

Weiland J, Davis W, Holter J, Mohammed J, Dorinsky P, Gadek J (1986) Lung neutrophils in the adult respiratory distress syndrome. Am Rev Respir Dis 133:218–225

Weissman C, Kemper M, Damask M, Ashkanazi J, Hyman A, Kinney J (1984) Effect of routine intensive care interactions on metabolic rate. Chest 86:815–818

Weissman C, Kemper M, Elwyn D, Ashkanazi J, Hyman A, Kinney J (1986) The energy expenditure of the mechanically ventilated critically ill patient: an analysis. Chest 89:254–259

Westaby S (1986) Mechanisms of membrane damage and surfactant depletion in acute lung injury. Intensive Care Med 12:2–5

Wise W, Cook J, Halushka P, Knapp D (1980) Protective effects of thromboxane synthetase inhibitors in rats in endotoxic shock. Circ Shock 46:854–859

Chapter 8

Right Ventricular Performance and Positive End-Expiratory Pressure Ventilation

H. Forst, J. Racenberg, K. Peter and K. Messmer

It is now well recognized that the application of positive end-expiratory pressure (PEEP) to the ventilation of patients with adult respiratory distress syndrome (ARDS) improves the arterial oxygen tension but may be associated with a fall in cardiac output, thus leading to the net effect of unchanged or even reduced systemic oxygen transport. However, the factors causing this decrease in cardiac output remain controversial (Craig et al. 1985).

Possible Mechanisms of the Haemodynamic Side Effects of PEEP

It is well known that a rise in pleural pressure and the corresponding increase in intra-cavitary right atrial pressure lead to a decrease in venous return by reducing driving pressure (Fig. 8.1). While a reduction in venous return is accepted as a factor contributing to the fall in cardiac output, the alternative concept of lung stretch-induced release of humoral substances which in turn depress myocardial function has not been substantiated (Manny et al. 1978; Liebman et al. 1978; Grindlinger et al. 1979).

Left Ventricular Performance

Most investigators reporting a depression of left ventricular (LV) function following the application of PEEP have used LV filling pressure to assess concomitant changes in LV preload (Scharf and Ingram 1977; Manny et al. 1978; Cassidy et al. 1978). As

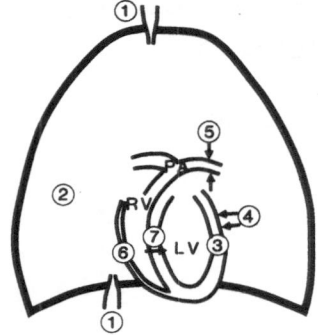

(1) Venous return

(2) Humoral factors

(3) Left ventricular contractility

(4) Ventricular tamponade

(5) Pulmonary vascular resistance

(6) Right ventricular contractility

(7) Ventricular interference

Fig. 8.1. Possible mechanisms (1–7) for the haemodynamic consequences of ventilation with positive end-expiratory pressure (PEEP). For details see text.

the left atrial or LV end-diastolic pressure increases or remains constant, impaired LV function is assumed to develop. Therefore all conclusions drawn depend upon the accurate measurement of changes in pressure between the pericardium and the surrounding lung. Oesophageal pressure underestimates intrathoracic pressure especially when higher levels of PEEP are applied (Prewitt and Wood 1979). The pressure in the surroundings of the heart is not distributed uniformly and rises to higher values than measured in other locations within the thoracic cavity (Fewell et al. 1980a; Marini et al. 1981). As a consequence, studies based on more accurate measurements of transmural pressure have failed to demonstrate evidence of impaired LV function (Scharf et al. 1979; Fewell et al. 1980b; Haynes et al. 1980; Calvin et al. 1981; Marini et al. 1981). The non-linear relationship between LV transmural pressure and LV volume makes the magnitude and direction of pressure and volume changes during PEEP a function of the ventricular filling conditions (Ditchey 1984).

Conclusions concerning the influence of an elevated LV end-diastolic pressure on LV contractility are usually based on the assumption that PEEP does not alter LV compliance. However, LV compliance has been found to decrease during PEEP ventilation without any evidence of impaired LV contractility (Scharf et al. 1979; Haynes et al. 1980; Santamore et al. 1984).

Right Ventricular Performance

In order to assess the importance of right ventricular (RV) dysfunction as a potential cause for the reduction in cardiac output associated with PEEP, those factors determining RV function will be discussed in detail with respect to how they are influenced by the application of PEEP (Table 8.1).

Table 8.1. Determinants of right ventricular performance, their approximations and additional factors which can be affected by PEEP

Preload (end-diastolic volume)
    Intracavitary filling pressure
    Transmural filling pressure
        Intrathoracic pressure
        Ventricular interference
        Pericardium

Afterload (wall stress)
    Pulmonary artery pressure
    Pulmonary vascular resistance
        Ventricular volume
        Wall thickness
        Intrathoracic pressure

Contractility
    RV free wall
        Myocardial perfusion
        Contraction pattern
    Septum
        LV function
        Ventricular interference

Preload

RV End-diastolic Volume and Filling Pressure. The best approximation of RV preload is RV end-diastolic volume, a parameter which is not easily available in patients (Matthay and Berger 1983). The measurement of ejection fraction by thermodilution techniques using a fast-response thermistor may be the solution to this problem (Kay et al. 1983). However, the RV is a crescent-shaped chamber surrounded by a concave free wall and a convex interventricular septum. Hence changes in chamber size do not necessarily reflect equivalent changes in tension within the RV wall. A more easily available approximation of preload is transmural pressure, which requires the measurement of intrathoracic pressure but does not take into account altered RV compliance.

Qvist et al. (1975) have demonstrated that the calculated transmural filling pressure decreases with the application of PEEP, but with adequate transfusion of blood the filling pressure can be returned to control level. RV and LV volumes assessed by the thermodilution technique (Fewell et al. 1980b) were markedly diminished during PEEP ventilation in dogs if the blood volume was not expanded at the same time. The changes in RV volume as determined by Laver et al. (1979) using radionuclide ventriculography differed according to whether PEEP was applied to a normal healthy subject or to patients with ARDS before or after volume replacement. In patients with ARDS both LV and RV volume indices assessed by scintigraphy decreased after the addition of PEEP without blood volume expansion (Viquerat et al. 1983).

Ventricular Interference. The pressure–volume relation during diastole and thus RV preload can be directly affected by alterations in both LV filling and the function of the LV free wall (Santamore et al. 1979; Molaug et al. 1981). The intact pericardium exerts a certain, although moderate, restrictive effect on cardiac chamber volume which predominantly influences RV filling (Tyson et al. 1984). Due to the different viscoelastic properties of the pericardium in humans as compared with dogs (Lee and Boughner 1985), the role of pericardial constraint might be more important in man. Both ventricles are characterized by a non-linear relationship between end-diastolic pressure and end-diastolic volume (Sibbald and Driedger 1983). This principle may account for difficulties in the interpretation of the results of studies in which preload was assessed solely by pressure measurements (Scharf and Ingram 1977; Manny et al. 1978; Cassidy et al. 1978; Prewitt and Wood 1979).

Afterload

Because of the pleomorphic shape and the considerable variability in configuration of the human RV, wall stress cannot simply be calculated by Laplace's law as is generally accepted for the LV. Furthermore, wall thickness, which is another component in the Laplace formula, cannot easily be determined in the RV.

Pulmonary Vascular Resistance. Widely used indices of RV afterload are mean pulmonary artery pressure and pulmonary vascular resistance. For convenience most authors use calculated pulmonary resistance when evaluating the afterload of the RV during PEEP ventilation. It should be realized, however, that the total load seen by the right heart during ejection is defined by both the DC and the AC impedance of the pulmonary artery bed (Piene and Sund 1979). While calculated vascular resistance reflects the relation between *mean* pressure and *mean* flow, the oscillatory part of impedance determines the relation between the pulsatile components of pressure and flow.

High levels of PEEP applied to normal lungs increase pulmonary pressure and pulmonary vascular resistance. However, this may not hold true for patients with acute respiratory failure. Pulmonary vascular resistance has been shown to be a function of lung volume, with its lowest value at functional residual capacity. Lung volumes greater than functional residual capacity are associated with an increased total pulmonary vascular resistance (Canada et al. 1982). RV systolic function is extremely sensitive to pressure loading (Sibbald et al. 1983). Pulmonary artery hypertension may be the consequence of underlying disease, as for example in patients with ARDS, or it may be additionally aggravated by the onset of ventilation with PEEP.

Dilatation of the thin-walled RV in patients with acute respiratory failure and positive pressure ventilation as reported by Laver et al. (1979) will further increase RV afterload. The enlargement of the RV increases wall stress and the local myocardial oxygen demand, particularly within the RV free wall, and this may thereby adversely affect RV contractility.

Contractility

Right Coronary Artery Blood Flow. The pattern of pulsatile right coronary artery blood flow at rest differs from the flow pattern in the left coronary artery. In contrast to the situation in the LV a significant fraction of pulsatile flow reaches the right myocardium during systole (Lowensohn et al. 1976; Bellamy and Lowensohn 1980). In dogs with RV hypertension coronary flow is characterized by a reduction in systolic flow and an increase in early diastolic flow (Lowensohn et al. 1976). This reduction in systolic flow is inversely related to the elevation of RV systolic pressure. However, a linear relation was found between aortic pressure and right coronary artery flow during diastole and systole when peak RV pressure was lower than systemic pressure (Bellamy and Lowensohn 1980). Nevertheless, when comparing the dynamics of right coronary blood flow in dogs with those in humans differences in the pattern of distribution between the coronary vessels have to be taken into account.

Myocardial Perfusion. Vlahakes et al. (1981) have demonstrated in dogs that with increasing RV afterload RV myocardial blood flow fails to increase in proportion to demand. However, if aortic pressure and thus myocardial perfusion are maintained by the infusion of phenylephrine, RV function is improved.

In contrast to these findings, moderate constriction of the pulmonary artery is associated with a significant increase in flow to the free wall of the RV and the interventricular septum (Gold and Bache 1982). In this study in dogs severe RV pressure overload was followed by RV subendocardial hypoperfusion, which also could be reversed by restoring aortic blood pressure. These findings support the conclusion of Brooks et al. (1971) that RV failure due to severe pulmonary artery obstruction can be reversed simply by increasing right coronary artery perfusion.

In ponies myocardial blood flow during acute RV systolic hypertension underwent an alteration in its distribution in favour of the RV free wall and the right part of the septum without any evidence of preferentially underperfused areas of myocardium (Manohar et al. 1978).

Myocardial Blood Flow during PEEP. Other investigators have measured myocardial perfusion and its distribution during PEEP ventilation using the microspheres method. Total myocardial blood flow was reduced along with a decrease in cardiac output when PEEP was applied to normal lungs (Manny et al. 1979; Fewell et al. 1980a); the same effects were observed after the induction of pulmonary oedema (Beyer and Messmer 1982). When taken as a fraction of cardiac output myocardial blood flow rose in favour of the RV, with no significant change in the subendocardial/ epicardial perfusion ration (Beyer and Messmer 1982).

Since it remains unknown whether myocardial perfusion during ventilation with PEEP meets the actual demands of the RV, one cannot exclude regional perfusion deficits which may lead to an impairment in local contractility of the RV free wall and thereby contribute to the decrease in cardiac output.

RV Contraction Pattern. An analysis of regional contraction patterns of the RV in dogs (Meier et al. 1980) has revealed that in systole the RV free wall undergoes a sequential contraction, which is initiated at the apex and terminates in the conus. The outflow tract, anatomically and embryologically a distinct region, contracts later and remains contracted longer than does the inflow tract (Raines et al. 1976). The myocardial segments of the RV free wall may remain isometric or even increase their length before the end of ejection (Pouleur et al. 1980). A significant part of RV ejection apparently occurs during relaxation of the RV free wall. Furthermore, RV contraction patterns seem to be sensitive to pharmacological intervention (Armour et al. 1970). The normal sequence of contraction from the sinus to the conus may be reversed by the infusion of noradrenaline. In this situation a marked intraventricular pressure gradient is created from sinus towards conus (Armour et al. 1970).

Care must be taken in the interpretation of some of these results, because many of these experiments were performed with both the pericardium and the thorax open—conditions which have been shown to influence the movements of both ventricular walls (Rushmer et al. 1952).

Ventricular Interference. The ventricles are usually considered as two pumps in series, but they also act as pumps in parallel. The common septum, together with a non-compliant pericardium, leads to ventricular interference. The fact that the contractile integrity of the RV free wall is not necessary for maintaining normal cardiac function, at least not under resting conditions, has led to the postulate by Brooks et al. (1971) that contraction of the intraventricular septum alone may be sufficient to maintain the pump function of the RV. In fact, the pressure signal obtained from the RV does not only reflect the pressure generated by the RV free wall, but includes a contribution from the pressure developed by contraction of the LV (Oboler et al. 1973; Feneley et al. 1985).

The concept of RV pump function being enhanced by the LV and septum has received some support from a study in pigs, in which the contractile response of the RV following infarction of three distinct areas of the LV was investigated (Brooks et al. 1977). The contractile parameters of the RV were significantly altered by ischaemia of the anteroseptal and inferoseptal areas, but not by ischaemia of the anterolateral wall. Thus RV performance was affected only when the septum was directly involved (Brooks et al. 1977).

Many studies have revealed changes in the LV pressure–volume relationship secondary to changes in RV volume. While some investigators note that this coupling is independent of an intact pericardium others stress that ventricular interaction is significantly affected by the presence or absence of an intact pericardium.

Laver et al. (1979) suggested that ventricular interference might play an important role in the therapy of acute respiratory failure in patients requiring mechanical ventilation with high airway pressures. They concluded that enthusiastic volume replacement in the presence of a restricted pulmonary vasculature may result in "LV end-diastolic tamponade". Recent echocardiographic findings from patients with ARDS (Jardin et al. 1981) indicate that PEEP can cause a leftward displacement of the interventricular septum. Similar findings have been observed in patients spontaneously breathing on CPAP (continuous positive airway pressure) (Jardin et al. 1984).

In contrast with these observations data obtained by sonomicrometry in conscious dogs do not support an abnormal shift of the interventricular septum at higher airway pressures (Rankin et al. 1982). Ventricular dimensions as assessed by means of radio-opaque markers and biplane cinefluoroscopy in dogs (Cassidy et al. 1982) revealed a PEEP-associated increase in the distance between the RV free wall and the septum, while the three LV dimensions measured decreased, indicating a reduction in calculated LV volume.

Sibbald et al. (1983) reported that patients presenting with ARDS, pulmonary artery hypertension and increased RV end-diastolic volume exhibited a significant alteration in LV diastolic mechanical properties. Viquerat et al. (1983), however, have not confirmed these findings in their study on patients with ARDS. Moreover, in a recent study in dogs Cassidy and Ramanthan (1984) found a similar displacement of the relative position of the septum into the LV as observed for the anterior or posterior wall. The authors thus concluded that deformation of the LV was more likely to be caused by a compression effect of the lungs than by dilatation of the RV.

In conclusion, on the basis of the literature reviewed the possible mechanisms of the haemodynamic side effects elicited by PEEP ventilation can be summarized as follows:

The primary effect of PEEP on the normal heart is characterized by an increase in RV afterload and a simultaneous decrease in RV preload.

As result of diminished transpulmonary blood flow LV end-diastolic volume becomes markedly reduced.

The results concerning the role of relative underperfusion of the RV free wall and the role of ventricular interference are somewhat controversial.

Studies on Myocardial Perfusion and Ventricular Interference

Our own group has been involved in the evaluation of the haemodynamic consequences of PEEP since 1982 (Beyer and Messmer 1982). The experimental investigations have comprised the analysis of the effect of PEEP on local and global contractility of the RV (Racenberg et al. 1983; Forst et al. 1984, 1986). The most recent studies were designed to answer the following questions:

Does the impairment of RV contractility secondary to myocardial ischaemia contribute to the haemodynamic consequences of PEEP?

Does PEEP shift the interventricular septum to the left as a result of augmented RV volume?

Methods

We have studied mongrel dogs, anaesthetized with pentobarbital and buprenophin and ventilated with a nitrous oxide/oxygen mixture on a volume-cycled respirator. Intraventricular pressures were obtained using micromanometer tip-catheters. Cardiac output was determined by the thermodilution method. Through a right thoracotomy and after pericardiotomy miniaturized ultrasonic transducers were

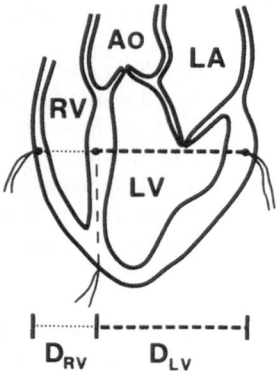

Fig. 8.2. Schematic drawing of the positioning of piezoelectric crystals in the free walls of both ventricles and within the interventricular septum. Right (D_{RV}) and left (D_{LV}) ventricular diameter are assessed by sonomicrometry. Ao, aorta; LA, left atrium.

implanted along the major axis within the RV free wall. Thus, shortening of the RV myocardial segment as well as the velocity of myocardial fibre shortening as parameters of local contractility could be measured by sonomicrometry (Racenberg et al. 1983).

In a special series of experiments septal motion was assessed by means of an ultrasonic crystal implanted through the tract of a 16-gauge needle within the interventricular septum, with corresponding crystals in the RV and LV free wall. This allowed simultaneous assessment of the diameters of both ventricles (Fig. 8.2). Intrathoracic pressure was measured by a micromanometer fixed at the right atrial level of the pericardial surface. The tip of the catheter was held within a silicon tube to protect it from motion artefacts. The lumen of the tube communicated freely with the pleural cavity via multiple side-holes. To induce ischaemia of the RV free wall the right coronary artery was permanently ligated. To assess the effect of pericardial constraint the pericardium was left either open or closed.

Autologous blood for volume substitution during PEEP was obtained by haemodiluting the animals isovolaemically with Dextran 60. The mean haematocrit was thereby reduced to 27% and remained constant during the experiment. After air-tight closure of the chest the remaining air was removed by a chest drain. The end-expiratory pressure was raised to 10, 15 and 20 cm H_2O and lowered to 10 and 0 respectively. PEEP levels of 10 and 20 cm H_2O were held constant for a period of 20 minutes. The RV transmural filling pressure—the difference between right atrial and intrathoracic pressure—was maintained constant during the PEEP intervention by blood transfusion.

Results

Contractility

Global contractility as derived from intraventricular pressure measurements (maximal velocity of intraventricular pressure rise, dp/dt_{max}, and maximal velocity of contractile element shortening, V_{max}) remained unchanged 5 and 30 minutes after occlusion of the right coronary artery, whereas *local* contractility (velocity of segment shortening and the degree of shortening) deteriorated. These data indicate that changes in local contractility cannot be detected when one measures exclusively the parameters of global contractility (dp/dt_{max}, V_{max}). During PEEP of 20 cm H_2O a further deterioration of the local contractility parameters was observed, again without concomitant changes in dp/dt_{max} and V_{max}.

Cardiac Output

The changes in cardiac output during different levels of PEEP are depicted in Fig. 8.3. With the pericardium left open and the right coronary artery intact (group I) cardiac output was unaffected by PEEP. When, however, the pericardium was closed (group II), cardiac output decreased with increasing PEEP. The comparison of groups II and IV, both consisting of animals with the pericardium closed, demonstrates that the effect of PEEP on cardiac output is not enhanced by the ligature of the right coronary artery. This does not hold true, however, when groups I and III, comprising animals with the pericardium left open, are compared.

An explanation for this finding may be that the increase in RV volume causes bulging of the RV free wall which is in part prevented when the pericardium is left intact or closed by sutures (Goto et al. 1985). It should be noted, however, that after infarction of the RV free wall maximal dilatation of the RV free wall occurred before the application of PEEP, i.e. PEEP was applied at maximal RV dilatation.

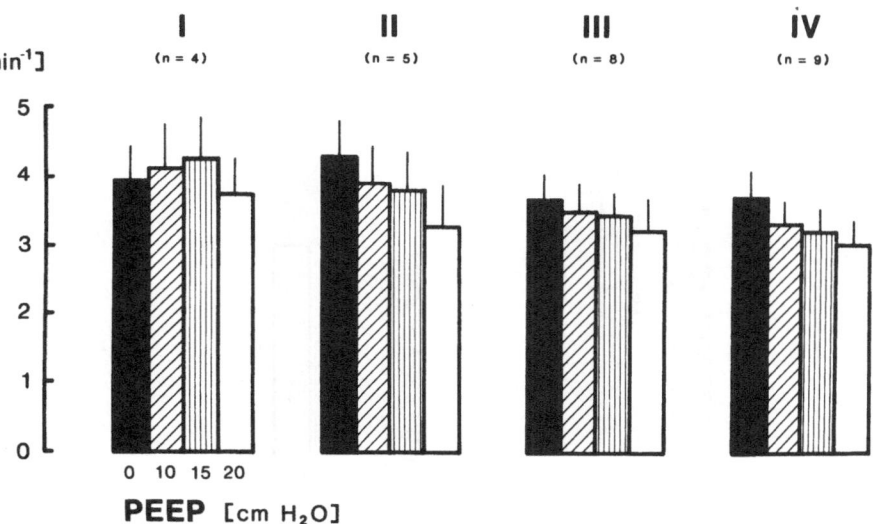

Fig. 8.3. Cardiac output of dogs ventilated with increasing levels of PEEP (10, 15 and 20 cm H_2O). RV filling pressure was kept constant by means of blood transfusion. Group I: pericardium open, right coronary artery (RCA) intact. Group II: pericardium closed, RCA intact. Group III: pericardium open, RCA ligated. Group IV: pericardium closed, RCA ligated. For further explanation see text. Values are mean ± SEM.

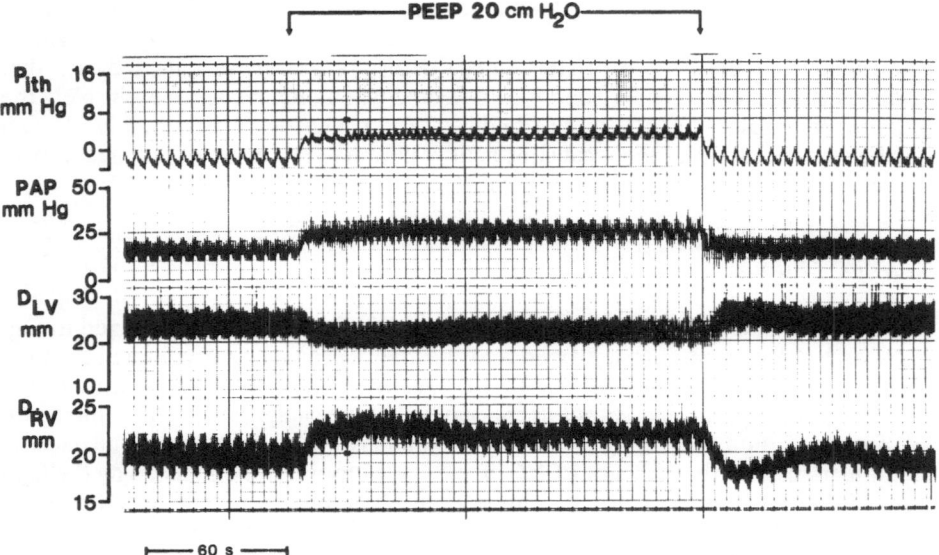

Fig. 8.4. Original recording of intrathoracic pressure (P_{ith}), pulmonary artery pressure (PAP), and left (D_{LV}) and right (D_{RV}) ventricular diameter obtained from a dog with open pericardium and intact RV free wall. Note the reversible increase in D_{RV} and the concomitant decrease in D_{LV} with the onset of PEEP.

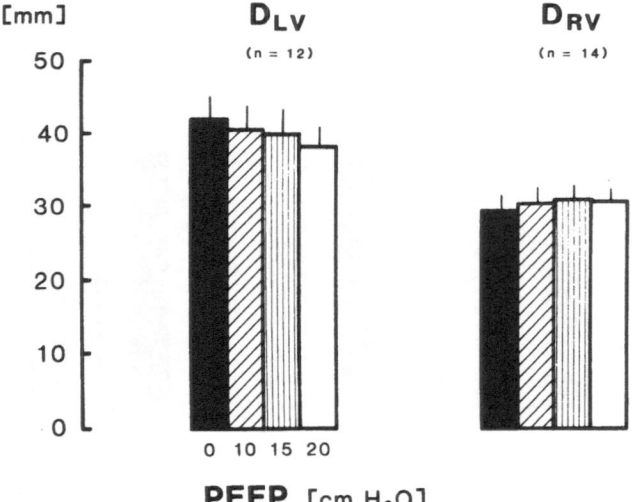

Fig. 8.5. Maximal values of left (D_{LV}) and right (D_{RV}) ventricular diameter assessed by sonomicrometry at increasing levels of PEEP (10, 15 and 20 cm H_2O). Note the stepwise decrease in D_{LV}, while mean D_{RV} reaches its maximum with PEEP at 15 cm H_2O. Diameter changes were independent of the condition of the pericardium and the RV free wall (see text). Values are mean ± SEM.

Ventricular Diameter

One possible mechanism for the augmentation of the RV volume is a shift of the intraventricular septum to the left. Fig. 8.4 shows an original recording from an experiment with simultaneous registration of the distance between interventricular septum and RV and LV free walls, respectively. Intrathoracic pressure and pulmonary artery pressure increase with the onset of PEEP, as does RV diameter; in contrast LV diameter decreases, a finding indicative of a shift of the interventricular septum to the left.

In all groups the maximal value of LV diameter decreased significantly with stepwise increases in PEEP, while the mean distance between septum and free wall of the RV reached its maximum at PEEP values higher than 10 cm H_2O (Fig. 8.5). Whereas the LV septal-to-free-wall distance shortened in all experiments independent of either ligation of the right coronary artery or closure of the pericardium, the changes in RV diameter were not uniform. In 10 of 14 animals RV diameter increased with PEEP, while in 4 dogs RV septal-to-free-wall distance decreased or remained unchanged. The increase in the septal-to-free-wall distance of the RV was not followed by a corresponding decrease in LV diameter, indicating a reduction in cardiac size during PEEP. This finding implies that a shift of the septum cannot be the sole reason for the changes observed in LV diameter. Relative hypovolaemia despite blood transfusion or compression of the heart by the lungs may account for these observations.

Position of the Interventricular Septum

The alterations in LV dynamic geometry associated with PEEP ventilation are illustrated by the LV transmural pressure–diameter loops constructed from single cardiac cycles during the same experiment (Fig. 8.6). Displacement of the loop to the left is again indicative of septal shift. Furthermore, during PEEP the elongation of the septal–free wall axis during isovolumic systole and a decrease during relaxation indicate abnormal septal motion. A "paradoxical" pattern of motion has been described previously as one of the characteristic echocardiographic features of selective RV volume overload (Pearlman et al. 1976). Tanaka et al. (1980) suggested that the diastolic shape and motion of the septum are determined by the intraventricular pressure gradient between the ventricles.

The trans-septal gradient of ventricular pressures in our study revealed a marked reduction during PEEP despite volume replacement, although we could not obtain a negative intraventricular pressure gradient between the LV and RV during diastole as reported by Tanaka et al. (1980).

In our study we were unable to measure all dimensions of the LV and their relative positions in the thoracic cavity. Therefore our data do not allow exclusion of a paradoxical movement of the LV lateral free wall corresponding to a LV tamponade by the lungs (Cassidy and Ramanathan 1984). However, a shift of the septum seems most likely: the increase in RV afterload induced by partial pulmonary artery occlusion is followed by a decrease in the LV septal–free wall axis exclusively, while the anterior–posterior and LV major axis from base to apex remain unaffected (Olsen et al. 1983). The transmural pressure–dimension loops obtained from animals with acute constriction of the pulmonary artery do reflect a rearrangement of LV geometry (Visner et al. 1983) which is very similar to the changes in ventricular geometry suggested by the loops from our experiments during PEEP.

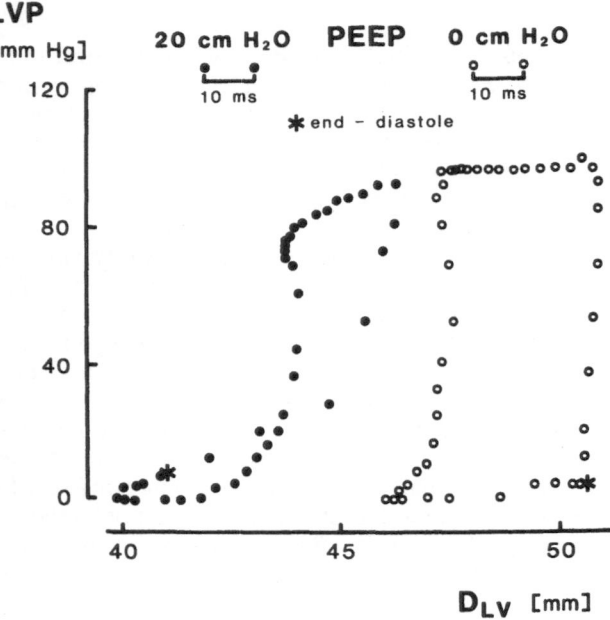

Fig. 8.6. Left ventricular pressure–diameter loops constructed from single cardiac cycles of an animal with open pericardium and intact right coronary artery. *Open circles* represent the pressure–diameter relation at 10 ms intervals during control conditions; *filled circles* depict data obtained during PEEP of 20 cm H_2O. While during control conditions the loop corresponds to the well-known pressure–volume curve, PEEP causes a rearrangement of ventricular geometry with increasing septal-to-free-wall distance (D_{LV}) during isovolumic systole and a decrease in D_{LV} during ventricular relaxation.

Conclusion

A review of the literature emphasizes the predominant role of reduced RV preload and simultaneously increased RV afterload in the haemodynamic consequences of PEEP ventilation. The relative importance of both RV dilatation and ventricular interference as contributing factors is still the subject of considerable controversy.

Our studies in dogs with normal lungs have shown that the haemodynamic side effects of PEEP after blood volume expansion are not influenced by a deterioration in local contractility of the ischaemic RV free wall.

Data on ventricular interference suggest that the increased RV septum-to-free-wall dimension after volume expansion results in a shift and a paradoxical movement of the interventricular septum which alters the dynamic geometry of the LV. This may be due to a reduction in the trans-septal pressure gradient. When both filling pressures are kept constant by blood volume expansion, ventricular interference appears as a major factor causing the haemodynamic side effects of PEEP.

This work was supported by DFG Sonderforschungsbereich 320.

References

Armour JA, Page JB, Randall WC (1970) Interrelationship of architecture and function of the right ventricle. Am J Physiol 218:174–179

Bellamy RF, Lowensohn HS (1980) Effect of systole on coronary pressure–flow relations in the right ventricle of the dog. Am J Physiol 238:H481–H486

Beyer J, Messmer K (1982) Organdurchblutung und Sauerstoffversorgung bei PEEP. Springer, Berlin Heidelberg New York (Anaesthesiologie und Intensivmedizin, vol 145)

Brooks H, Kirk ES, Vokonas PS, Urschel CW, Sonnenblick EH (1971) Performance of the right ventricle under stress: relation to right coronary flow. J Clin Invest 50:2176–2183

Brooks H, Holland R, Al-Sadir J (1977) Right ventricular performance during ischemia: an anatomic and hemodynamic analysis. Am J Physiol 233:H500–H513

Calvin JE, Driedger AA, Sibbald WJ (1981) Positive end-expiratory pressure (PEEP) does not depress left ventricular function in patients with pulmonary edema. Am Rev Respir Dis 124:121–128

Canada E, Benumof JL, Tousdale FR (1982) Pulmonary vascular resistance correlates in intact normal and abnormal canine lungs. Crit Care Med 10:719–723

Cassidy SS, Ramanthan M (1984) Dimensional analysis of the left ventricle during PEEP: relative septal and lateral wall displacements. Am J Physiol 246:H792–H805

Cassidy SS, Robertson CH, Pierce AK, Johnson RL (1978) Cardiovascular effects of positive end-expiratory pressure in dogs. J Appl Physiol 44:743–750

Cassidy SS, Mitchell JH, Johnson RL (1982) Dimensional analysis of right and left ventricles during positive-pressure ventilation in dogs. Am J Physiol 242:H549–H556

Craig KC, Pierson DJ, Carrico CJ (1985) The clinical application of positive end-expiratory pressure (PEEP) in the adult respiratory distress syndrome (ARDS). Resp Care 30:184–201

Ditchey RV (1984) Volume-dependent effects of positive airway pressure on intracavitary left ventricular end-diastolic pressure. Circulation 69:815–821

Feneley MP, Gavaghan TP, Baron DW, Branson JA, Roy PR, Morgan JJ (1985) Contribution of left ventricular contraction to right ventricular systolic pressure in the human heart. Circulation 71:473–480

Fewell JE, Abendschein DR, Carlson CJ, Rapaport E, Murray JF (1980a) Mechanism of decreased right and left ventricular end-diastolic volumes during continuous positive-pressure ventilation in dogs. Circ Res 47:467–472

Fewell JE, Abendschein DR, Carlson CJ, Murray JF, Rapaport E (1980b) Continuous positive-pressure ventilation decreases right and left ventricular end-diastolic volumes in the dog. Circ Res 46:125–132

Forst H, Racenberg J, Fujita Y, Zeintl H, Messmer K (1984) Does "PEEP" influence right ventricular performance? Eur Surg Res 16 [Suppl 1]:34

Forst H, Racenberg J, Peter K, Messmer K (1986) The right ventricle: a pump or a reservoir? Eur J Anaesthesiol 3:74

Gold FL, Bache RJ (1982) Transmural right ventricular blood flow during acute pulmonary artery hypertension in the sedated dog. Circ Res 51:196–204

Goto Y, Yamamoto J, Saito M et al. (1985) Effects of right ventricular ischemia on left ventricular geometry and the end-diastolic pressure–volume relationship in the dog. Circulation 72:1104–1114

Grindlinger GA, Manny J, Justice R, Dunham B, Shepro D, Hechtman HB (1979) Presence of negative inotropic agents in canine plasma during positive end-expiratory pressure. Circ Res 45:460–467

Haynes JB, Carson SD, Whitney WP, Zerbe GO, Hyers TM, Steele P (1980) Positive end-expiratory pressure shifts left ventricular diastolic pressure–area curves. J Appl Physiol 48:670–676

Jardin F, Farcot JCh, Boisante L, Curien N, Margairaz A, Bourdarias J-P (1981) Influence of positive end-expiratory pressure on left ventricular performance. N Engl J Med 304:387–392

Jardin F, Farcot JCh, Gueret P, Prost JF, Ozier Y, Bourdarias JP (1984) Echocardiographic evaluation of ventricles during continuous positive airway pressure breathing. J Appl Physiol 56:619–627

Kay HR, Afshari M, Barash P et al. (1983) Measurement of ejection fraction by thermal dilution techniques. J Surg Res 34:337–346

Laver MB, Strauss HW, Pohost GM (1979) Right and left ventricular geometry: adjustments during acute respiratory failure. Crit Care Med 7:509–519

Lee JM, Boughner DR (1985) Mechanical properties of human pericardium: differences in viscoelastic response when compared with canine pericardium. Circ Res 55:475–481

Liebman PR, Patten MT, Manny J, Shepro D, Hechtman HB (1978) The mechanism of depressed cardiac output on positive end-expiratory pressure (PEEP). Surgery 83:594–598

Lowensohn HS, Khouri EM, Gregg DE, Pyle RL, Patterson RE (1976) Phasic right coronary artery blood flow in conscious dogs with normal and elevated right coronary pressures. Circ Res 39:760–766

Manny J, Patten MT, Liebman PR, Hechtman HB (1978) The association of lung distension, PEEP and biventricular failure. Ann Surg 18:151–157

Manny J, Justice R, Hechtman HB (1979) Abnormalities in organ blood flow and its distribution during positive end-expiratory pressure. Surgery 85:425–432

Manohar M, Bisgard GE, Bullard V, Will JA, Anderson D, Rankin JH (1978) Myocardial perfusion and function during acute right ventricular systolic hypertension. Am J Physiol 235:H628–H636

Marini JJ, Culver BH, Butler J (1981) Effect of positive end-expiratory pressure on canine ventricular function curves. J Appl Physiol 51:1367–1374

Matthay RA, Berger HJ (1983) Noninvasive assessment of right and left ventricular function in acute and chronic respiratory failure. Crit Care Med 11:329–338

Meier GD, Bove AA, Santamore WP, Lynch PR (1980) Contractile function in the canine right ventricle. Am J Physiol 239:H794–H804

Molaug M, Stokland O, Ilebeek A, Lekven J, Kiil F (1981) Myocardial function of the interventricular septum: effects of right and left ventricular pressure loading before and after pericardiotomy in dogs. Circ Res 49:52–61

Oboler AA, Keefe JF, Gaasch WH, Banas JS, Levine HJ (1973) Influence of left ventricular isovolumic pressure upon right ventricular pressure transients. Cardiology 58:32–44

Olsen CO, Tyson GS, Maier GW, Spratt JA, Davis JW, Rankin JS (1983) Dynamic ventricular interaction in the conscious dog. Circ Res 52:85–104

Pearlman AS, Clark CE, Henry WL, Morganroth J, Itscoitz SB, Epstein SE (1976) Determinants of ventricular septal motion: influence of relative right and left ventricular size. Circulation 54:83–91

Piene H, Sund T (1979) Flow and power output of right ventricle facing load with variable input impedance. Am J Physiol 237:H125–H130

Pouleur H, Lefevre J, Van Mechelen H, Charlier AA (1980) Free-wall shortening and relaxation during ejection in the canine right ventricle. Am J Physiol 239:H601–H613

Prewitt RM, Wood LDH (1979) Effect of positive end-expiratory pressure on ventricular function in dogs. Am J Physiol 236:H534–H544

Qvist J, Pontoppidan H, Wilson RS, Lowenstein E, Laver MB (1975) Hemodynamic responses to mechanical ventilation with PEEP. Anesthesiology 42:45–55

Raines RA, LeWinter MM, Covell JW (1979) Regional shortening patterns in canine right ventricle. Am J Physiol 231:1395–1400

Racenberg J, Fujita Y, Forst H, Brückner UB, Messmer K (1983) Rechtsventrikuläre Kontraktilität bei PEEP-Beatmung. Anaesthesist 32 [Suppl]:355–356

Rankin JS, Olsen CO, Arentzen CE et al. (1982) The effect of airway pressure on cardiac function in intact dogs and man. Circulation 66:108–120

Rushmer RF, Crystal DK, Wagner C (1952) The functional anatomy of ventricular contraction. Circ Res 1:162–170

Santamore WP, Meier GD, Bove AA (1979) Effects of hemodynamic alterations on wall motion in the canine right ventricle. Am J Physiol 5:H254–H262

Santamore WP, Bove AA, Heckman JL (1984) Right and left ventricular pressure–volume response to positive end-expiratory pressure. Am J Physiol 246:H114–H119

Scharf SM, Ingram RH (1977) Effects of decreasing lung compliance with oleic acid on the cardiovascular response to PEEP. Am J Physiol 233:H635–H641

Scharf SM, Brown R, Saunders N, Green LH, Ingram RH (1979) Changes in canine left ventricular size and configuration with positive end-expiratory pressure. Circ Res 44:672–678

Sibbald WJ, Driedger AA (1983) Right ventricular function in acute disease states: pathophysiologic considerations. Crit Care Med 11:339–345

Sibbald WJ, Driedger AA, Myers ML, Short AI, Wells GA (1983) Biventricular function in the adult respiratory distress syndrome: hemodynamic and radionuclide assessment, with special emphasis on right ventricular function. Chest 84:126–134

Tanaka H, Tei C, Nakao S, Tahara M, Sakurai S, Kashima T, Kanehisa T (1980) Diastolic bulging of the interventricular septum toward the left ventricle: an echocardiographic manifestation of negative interventricular pressure gradient between left and right ventricles during diastole. Circulation 62:558–563

Tyson GS, Maier GW, Olsen CO, Davis JW, Rankin JS (1984) Pericardial influences on ventricular filling in the conscious dog: an analysis based on pericardial pressure. Circ Res 54:173–184

Viquerat CE, Righetti A, Suter PM (1983) Biventricular volumes and function in patients with adult respiratory distress syndrome ventilated with PEEP. Chest 83:509–514

Visner MS, Arentzen CE, O'Connor MJ, Larson EV, Anderson RW (1983) Alterations in left ventricular three-dimensional dynamic geometry and systolic function during acute right ventricular hypertension in the conscious dog. Circulation 67:353–365

Vlahakes GJ, Turley K, Hoffman JE (1981) The pathophysiology of failure in acute right ventricular hypertension: hemodynamic and biochemical correlations. Circulation 63:87–95

Section III

Some Aspects of Ventilatory Support

Chapter 9

The Physiological Basis of Ventilatory and Respiratory Support

W. Kox

The indications for support in acute respiratory failure are based on clinical, radiological and laboratory evidence: dyspnoea, tachypnoea and panlobar alveolar infiltrates of one or both lungs accompanied by severe hypoxaemia with an arterial oxygen tension (PaO_2) or less than 75 mmHg when breathing a fraction of inspired oxygen (FiO_2) of more than 0.5 (Artigas et al. 1985). These clinical signs often require an immediate response: in an attempt to improve arterial oxygenation, continuous positive airways pressure (CPAP) is applied via a facemask or an endotracheal tube, but frequently some sort of mechanical ventilation is required. However, the mode of mechanical ventilation depends essentially upon a process of trial and error in which success is monitored by blood gas analysis. Although reliable, this empirical method does not disclose where the failure or potential for improvement in ventilation lies. A method by which the optimal ventilatory pattern could be found should encompass the different anatomical levels of the lungs and their function as well as their mechanics.

The objective in treating patients with varying forms of respiratory failure is to normalize lung function. This can be achieved by:

1. Normalizing alveolar ventilation
2. Maintaining pulmonary blood flow
3. İmproving the distribution and matching of ventilation with perfusion throughout all areas of the lungs
4. Reducing the work of breathing, and finally
5. Successfully discontinuing the respiratory and ventilatory support.

It therefore seems essential to assess independently lung function in respect of these five points and adjust the pattern of ventilation accordingly.

The Assessment of Lung Function

Inspired air containing 20.93% oxygen is transported down the airways to the blood in four distinct physical steps:

1. Active movement of the rib cage and diaphragm which creates a pressure gradient between the alveoli and the ambient air
2. Convective flow down the conducting airways along this pressure gradient
3. Molecular diffusion in the gas phase in the respiratory airways along partial pressure gradients, and
4. Molecular diffusion in the tissue phase across the alveolar–capillary membranes.

Lung Mechanics

Under normal physiological conditions gas exchange is initiated by upward movement of the rib cage and the contraction and hence downward movement of the diaphragm which thereby increases intrathoracic volume. The sudden increase in volume, which is transmitted to the lung periphery via the parietal and visceral pleura, creates a pressure difference between the alveoli and the ambient air. The negative intrapulmonary pressure during the initial phase of inspiration will cause air to flow first rapidly into the lungs and then less rapidly as alveolar pressure approaches ambient pressure (760 mmHg standard barometric pressure). During expiration the elastic recoil of the lungs suspended in the pleural cavity and the thoracic wall will exert positive pressure on the alveoli thereby squeezing gas towards the trachea, the vocal cords and the pharynx. The vocal cords offer the highest resistance to gas flow in normal healthy subjects. This effect may be enhanced when the vocal cords are intermittently closed whilst speaking.

The amount of air inspired and gas expired obviously depends not only on the pressure difference generated but also on the resistance of the conducting airways to gas flow and the compliance of the lung tissues. The relationship between these variables is best described by a pressure–volume loop the area of which represents the work of breathing (Fig. 9.1). The overall work of breathing (W_{RS}) can be divided into the elastic work (W_{el}) due to changes in lung compliance and the viscous work (W_{vis}) describing the work required to overcome the viscous resistance of the conducting airways to flow. Since all work of breathing consists of two components—the work exerted by the lungs (W_L) and that by the thoracic wall (W_W)—the relationship can be expressed by the equation:

$$W_{RS} = W_W + W_L = (W_{W.vis} + W_{W.el}) + (W_{L.vis} + W_{L.el})$$

W_{RS} in a healthy subject at rest constitutes only 1% of the total energy expenditure, and is almost entirely spent on the inspiratory chest wall movement. The elastic work of the lungs is the energy stored in the compliant lung tissues during inspiration, and is used to overcome the resistance to gas flow during expiration. The elastic work depends on the dynamic lung compliance ($C_{L.dyn}$) which is the change in volume per change in pressure during inspiration and expiration:

$$C_{L.dyn} = \Delta V/\Delta P$$

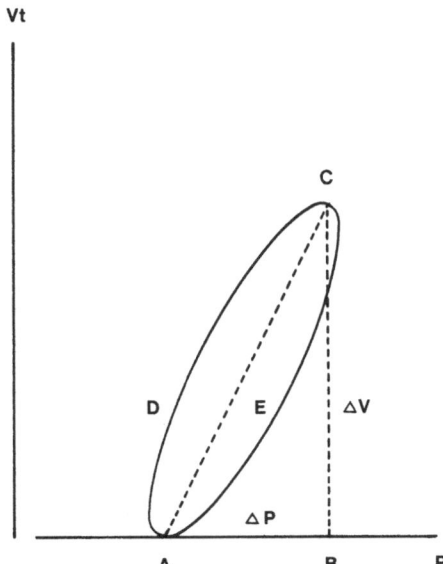

Fig. 9.1. Pressure–volume loop, where the area ABCA characterizes $W_{L,el}$, the area AECDA $W_{L,vis}$, ABCDA $W_{L.tot}$ and $\Delta V/\Delta P$ the dynamic lung compliance.

When ventilation is increased due to physical labour or septicaemia dynamic lung compliance becomes dependent on respiratory rate and tidal volume (Vt), which in turn will have a bearing on $W_{L.vis}$ and can therefore be standardized for 1 litre of tidal volume. It is then called the specific work of breathing:

$$sW_{L.vis} = (\phi Vt \times d\,P_{tp})/Vt$$

where the term in the brackets describes the area within the pressure–volume loop.

Convective Flow in the Conducting Airways

When an inspirate of pure oxygen is considered, convective ventilation results in a linear gas velocity in the trachea, and as more branches of the bronchial tree are traversed this velocity progressively decreases until at the alveolar wall velocity becomes zero. Simultaneously nitrogen and carbon dioxide molecules resident in the lung diffuse into the inspired oxygen, and the rate of upward diffusion may be represented by a linear gas velocity.

The relative contribution of the convective velocity to diffusive characteristics in an airway is given by the Peclet number, N_{Pe}. The Peclet number decreases by a factor of 10^5 from the trachea to the terminal bronchiole, which characterizes the great increase in cross-sectional area, and the decrease in diameter in successive branching generations. A critical zone is established between the convective velocity downwards and the diffusive velocity upwards at which the Peclet number equals unity. The interface of concentration between the oxygen molecules and the nitrogen molecules becomes stationary, that is a "stationary interface" is established (Cumming et al. 1971; Engel et al. 1973). The volume of gas proximal to the stationary interface is in the conducting airways, the so-called anatomical dead space. This phrase was coined because the first measurement of the volume of the conducting airways was done using an anatomical human cast by Loewy (1894). He found the volume to be 144 ml. The first difficulty with this method is to define the space in a region where volume is increasing rapidly with linear distance. Secondly, the dead space of the conducting airways can be changed by many factors and is not only anatomically but also functionally dependent.

Bowes et al. (1982) described an increase in dead space with an increase in either inspiratory or expiratory flow. They also found by model analysis that an increased inspired volume causes anatomical dead space to increase by two main mechanisms: the movement of the interface down towards the lungs during inspiration and the expansion of the airways during increased inspiration. Dead space was also found to be sensitive to inspiratory time and breathing pattern. If inspiratory time is prolonged the interface moves back up towards the trachea thus reducing dead space. This is most clearly shown by breath-holding. In contrast a high flow at end-inspiration will cause the interface to be established further down the lung. Anatomical dead space also depends on drugs acting on the bronchial muscles. Kox et al. (1982) found the volume of dead space significantly increased after the administration of atropine. Adrenaline, isoprenaline and salbutamol are known to have the same effect, whereas methocoline causes bronchoconstriction and therefore decreases dead space.

Anatomical dead space can be measured using a method described by Fowler in 1948. This uses a nitrogen meter or rapid gas analyser to give a functional assessment of the expired concentration of nitrogen following an inspirate of pure oxygen. The plot obtained consists of three parts (Fig. 9.2): the first phase contains no nitrogen and represents purely dead space gas; the second phase contains a varying mixture of dead space gas and alveolar gas; and the third phase, which is called the alveolar plateau or slope, represents the alveolar gas. Cumming and Guyatt (1982) determined anatomical dead space by utilizing a cumulative plot of expired gas volume (nitrogen) against total expired volume. They applied a second-order polynomial

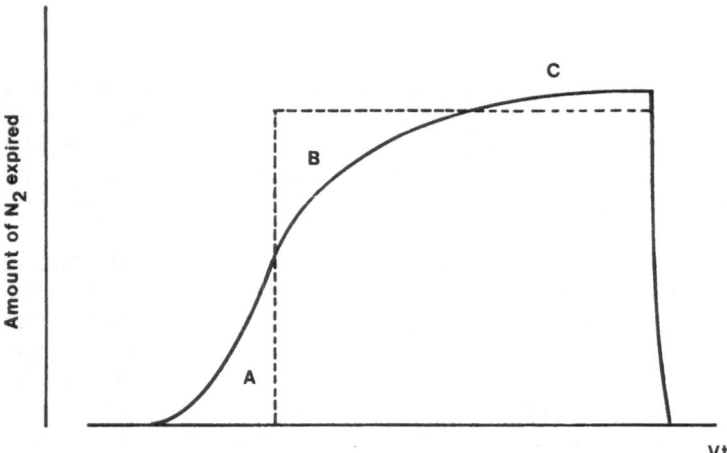

Fig. 9.2.   Fowler method of determining anatomical dead space, which is the volume of air defined by the broken line where area A is equal to area B. C is the alveolar slope.

regression to the data and an intercept was projected onto the abscissa at a volume which represented the dead space. Finally in a plot of expired gas concentration against expired volume, the second phase of this "expirogram" can be differentiated in order to find the greatest rate of change on the S-shaped curve. The projection of this point onto the abscissa will then yield dead space volume (Lanczos 1957).

Diffusive Mixing

The third phase of the expired concentration curve, the alveolar slope, has been explained by an uneven distribution of inspired gas followed by an expiration which differs in its time course from different regions. Krogh and Lindhard (1917) observed that the end-expired concentration of carbon dioxide was higher in the alveolar region than at the mouth and concluded that diffusive mixing was not absolutely uniform. They proposed that gas diffusion along each pulmonary pathway was incomplete over the time of a respiratory cycle, resulting in the most distal alveoli receiving less fresh inspired gas than the more proximal ones. These concentration differences would be arranged in sequence, which has subsequently led to the terms "stratified" or "series inhomogeneity". An alternative explanation for the difference in end-expiratory concentrations would be that mixing is perfect but unequal amounts of inspired gas enter different lung units, or that the initial volumes of the lung units differ. This distributional concentration difference has been termed "regional" or "parallel inhomogeneity".

However, the cause of this failure of complete mixing is still ill-understood, but might be due to mechanisms operating at one or more of three anatomical levels in the airways: (1) inequality of regional or lobar distribution of inspired gas; (2) inequality of distribution within lobes (*Pendelluft*); (3) failure of diffusive mixing within acini, perhaps due to the asymmetry of the anatomy.

The space in which gas mixing takes place can best be assessed by the measurement of functional residual capacity (FRC). As the ambient air or inspired gas moves down the conducting airways it becomes increasingly saturated with water vapour with a partial pressure up to 47 mmHg. When a stationary interface is established the water-vapour-saturated fresh gas will start mixing with the residual gas containing not only oxygen and nitrogen but also carbon dioxide evolving from the mixed venous blood in the pulmonary circulation. The alveolar partial pressure can now be derived and consequently the alveolar–arterial oxygen difference $D(A-a)O_2$ calculated:

$$D(A-a)O_2 = FiO_2 = (Pb - 47) - PaCO_2/0.8 - PaO_2$$

where 0.8 represents the respiratory quotient (Artigas et al. 1985) and Pb the measured barometric pressure. The $D(A-a)O_2$ is a reliable parameter for estimating the degree of respiratory failure. If the $D(A-a)O_2$ is greater than 350 mmHg some means of increasing the alveolar gas exchange of oxygen is required. This can be achieved in three different ways or, if necessary, by a combination of all three:

1. A further increase in FiO_2
2. The application of positive end-expiratory pressure (PEEP)
3. Inversed inspiratory to expiratory time (I:E) ratio.

An increase in FiO_2 will automatically lead to a higher alveolar partial pressure of oxygen thereby increasing alveolar oxygen tension. However, higher concentrations of oxygen are known to be toxic, damage alveolar epithelium and therefore lead to progressive alveolar collapse (Fischer 1980). The application of PEEP or CPAP in spontaneously breathing patients will increase FRC thereby making more surface area available for gas exchange. The indiscriminate use of PEEP or CPAP, however, can induce alveolar damage with consequent collapse and barotrauma. The use of inversed I : E ratios and their effect on FRC will be discussed later.

In order to avoid the adverse effects of PEEP or CPAP an accurate and repeatable test for measuring FRC is required. Recently, the multiple-breath nitrogen washout technique has been described as a method for the assessment of FRC in ventilated patients (Ozanne et al. 1981; Paloski et al. 1981; Ibancz et al. 1983). The principle of inert gas washout using mass balance enables a check to be made of the total quantity of nitrogen recovered from the lungs, thus ascribing a value to FRC. The pattern by which this nitrogen is recovered provides a figure for the diffusive mixing of the fresh with the residual gas, thus assessing alveolar gas mixing.

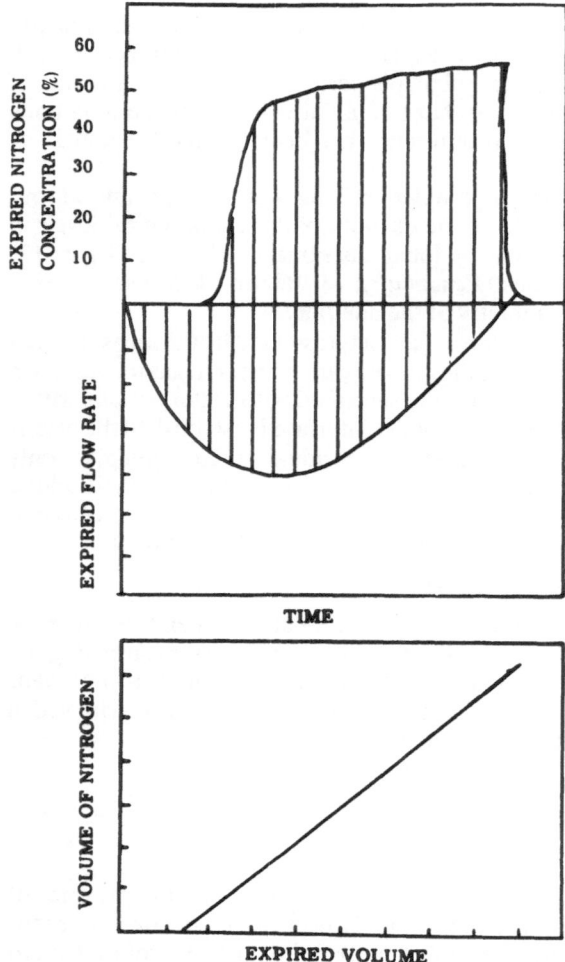

Fig. 9.3.   Flow–concentration plot sampled every 20 ms (vertical lines).

When a nitrogen-analyser or mass spectrometer is available, gas analysis and flow measurement with a Fleisch pneumotachograph can be passed to an on-line computer. The product of concentration and flow can then be calculated and will yield the nitrogen expired after the inspiration of a nitrogen-free gas (either argon–oxygen or pure oxygen) (Fig. 9.3). This method is based on the fact that the nitrogen in the accessible gas in the lungs (FRC) is replaced by argon (or oxygen). Thus the expired amount of nitrogen represents a percentage of accessible gas depending on the FiO_2; when this figure is multiplied by 100 over the percentage of nitrogen in the inspirate before the washout manoeuvre it yields the value for FRC. It is then possible to calculate the efficiency with which the nitrogen and the nitrogen-free gas mix in the gas phase. This is obtained from a logarithmic plot of the nitrogen remaining in the lungs against turnover, which is the sum of the tidal volumes (Vt) necessary to replace the volume of FRC (ΣVt/FRC). A second line can be plotted breath by breath, the slope of which represents the line of perfect mixing derived from the equation:

$$C_n = C_{n-1} \times (FRC/FRC + Vt)$$

where Vt is the tidal volume, C the concentration of nitrogen and n the breath number. The ratio of the intercepts of the two lines with any given volume of nitrogen, i.e. the turnover from the perfect mixing curve divided by the turnover from the experimental curve, multiplied by 100, gives the ventilatory efficiency of the accessible gas (Fig. 9.4).

In contrast, the inefficient part expressed as a percentage of the mean tidal volume minus the anatomical dead space as derived from an expirogram is called the alveolar dead space. When this volume is expressed as a percentage of the volume which is well mixed in the alveoli, an index of diffusive mixing in the alveoli, the alveolar gas mixing efficiency, is obtained. Changes in alveolar dead space (Vd) ventilation are generally believed to be reflected in the Vd/Vt ratio. The Vd/Vt ratio is based on the assumption that the anatomical dead space is constant. Since this is not the case it is imperative to be able to measure anatomical and alveolar dead space independently, as described in the above method.

Pulmonary Blood Flow

In erect man blood flow in the lungs decreases rapidly, from bottom to top, reaching very low values at the apex. This pattern is, of course, affected by posture and exercise. Since the total output of the right side of the heart goes through the lungs the cardiac output is equal to pulmonary blood flow. Under normal physiological conditions the ratio of alveolar ventilation to pulmonary blood flow is approximately 1 : 1 (West 1979). When a patient lies in the supine position the apical and basal flows become the same, but the posterior, i.e. the dependent part of the lung, has a higher blood flow than the anterior region. In the lateral position the dependent regions are again the best perfused.

The uneven distribution of blood flow is due to the hydrostatic pressure differences within the lungs. West (1982) distinguishes three zones:

Zone 1 is at the top or independent part of the lung where pulmonary arterial pressure is less than alveolar pressure (normally atmospheric) caused by the hydrostatic gradient within the pulmonary arterial tree. If arterial pressure is less than alveolar, this part of the lung is not perfused because the pulmonary capillaries are directly exposed to alveolar pressure and collapse when the pressure outside, i.e. the alveo-

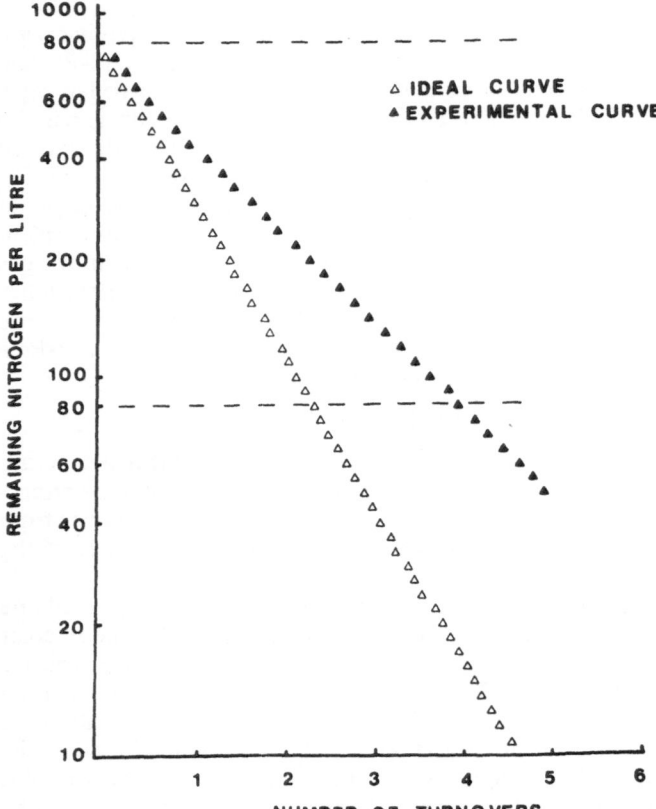

Fig. 9.4. Semilogarithmic nitrogen decay curve.

lar, exceeds the pressure inside. If the arterial pressure is reduced, as in hypovolaemic or septicaemic shock, or if the alveolar pressure is raised, as in positive pressure ventilation and the application of PEEP, then a non-perfused zone may be present.

Zone 2 is where pulmonary artery pressure exceeds alveolar pressure and alveolar exceeds venous pressure, and blood flow here depends on the difference between arterial and alveolar pressure. Since arterial pressure increases in this zone the more dependent the lung becomes, blood flow will increase accordingly, provided alveolar pressure remains constant.

Zone 3 is where venous pressure is greater than alveolar pressure. The thin-walled capillaries are thus held open and flow is determined by the arterial–venous pressure difference. Blood flow increases in this zone because of the increasing hydrostatic pressure and the constant alveolar pressure. However, at lower FRC levels blood flow is reduced in this zone probably because the larger pulmonary blood vessels in this area are held open by radial traction of the surrounding parenchyma, and collapse when poorly expanded.

The sequelae of shock and septicaemia can obviously have a detrimental effect on the delicate balance of ventilation and pulmonary blood flow. The sequestration of leucocytes and microemboli in the lung capillaries as described in previous chapters may lead to a far greater degree of uneven ventilation and blood flow than in the normal lung and the ventilation–perfusion inequality on gas transfer may be very severe.

This inequality can be assessed by calculating the venous admixture or shunt fraction ($\dot{Q}s/\dot{Q}t$) characterizing the blood which passes through capillaries which are not in contact with ventilated alveoli. Unfortunately, the shunt fraction does not reveal any information on the blockage of pulmonary capillaries. However, the accurate estimation of shunt fraction requires mixed venous blood sampling from the pulmonary artery. Since Swan–Ganz catheterization is necessary the measurement of the pulmonary artery pressures will give an idea of the degree of capillary occlusion. Shunt fraction is calculated by the equation:

$$\dot{Q}s/\dot{Q}t = \frac{CcO_2 - CaO_2}{CcO_2 - C\bar{v}rO_2} \, (\%)$$

CaO_2 is the arterial oxygen content, which can be derived from the arterial PO_2, pH and haemoglobin dissociation curve. $C\bar{v}O_2$ can similarly be calculated from a mixed venous blood sample (Severinghaus 1966). CcO_2 is the oxygen content of the blood which has just passed through the capillary bed and taken part in ideal gas exchange. It is calculated as the content with an arterial PO_2 equal to the ideal alveolar oxygen tension (derived from the alveolar gas equation) and a pH equal to the arterial pH.

A more accurate technique for measuring ventilation–perfusion inequality is the injection of a mixture of six gases of different solubilities. The concentration of the gases in the arterial blood and the expired gas are then measured by gas chromatography. From these data it is possible to compute ventilation–perfusion ratios continuously. Whereas normal lungs show a narrow distribution with little dispersion, in the presence of lung disease the distribution generally broadens and different diseases have characteristic distributions (Wagner et al. 1974).

Towards Best Ventilation

The ability to monitor the sequential steps of gas exchange should introduce a more critical approach to the use of the different modes of ventilation. Intermittent positive pressure ventilation (IPPV), although the most common treatment in acute respiratory failure, has a number of disadvantages, such as the increase in intrathoracic pressure which leads to a decrease in cardiac output and hence to a diminution in pulmonary blood flow. During IPPV a positive pressure impulse applied to an endotracheal tube is transmitted via the bronchial tree to the alveoli. Gas flow will follow the pathway of least resistance, thus favouring areas already inflated but avoiding areas which are blocked off already because of sputum retention, resulting in progressive atelectasis. A decrease in FRC in the supine position is also the result of an altered distribution of ventilation. As a consequence of this decrease the dependent areas move to a lower, flatter part of the volume–pressure curve where the compliance is low compared with that in the non-dependent regions.

The decrease in ventilation of the dependent regions during IPPV is also due to changes in diaphragmatic displacement. During IPPV the diaphragm is pushed caudally by the uniformly distributed pressure from the lungs which is opposed by a non-uniform, gravity-dependent hydrostatic pressure gradient from the abdominal contents. The diaphragmatic displacement is, therefore, greater in non-dependent areas and hence these areas will be ventilated preferentially. Since perfusion favours the dependent regions ventilation–perfusion mismatch will occur. The application of PEEP can reduce the preference of ventilation for the non-dependent areas by increasing the FRC and thereby increasing the compliance of the dependent lung. The non-dependent zones are shifted to the still higher but flatter part of the volume–pressure loop (Hillman 1986). The decrease in airway closure and atelectasis by PEEP lead to a better ventilation–perfusion match and reduced intrapulmonary shunt with an improvement in arterial PO_2. In non-homogeneous lung disease, however, higher levels of PEEP may lead to overdistension of normal alveoli, which causes a decrease in perfusion of those alveoli because of capillary compression. The result is an increase in shunt fraction as blood is diverted to diseased, non-ventilated areas. The concept of "optimum PEEP" was first described by Suter et al. (1975); best PEEP is tailored to recruit as many alveoli as possible while avoiding overdistension. Overall pulmonary compliance will then be maximal and coincide with the highest oxygen delivery despite a slightly reduced cardiac output.

Inversed I : E ratios and Alternative Inspiratory Flow Pattern

Under normal physiological conditions a healthy subject breathes with a respiratory rate of 10–20 per minute and a tidal volume of 400–800 ml depending on weight, height, posture and mental state. The inspiratory to expiratory time (I : E) ratio under these circumstances is 1 : 1.5, resulting in a rapid inspiratory and a lower expiratory flow which allows more time for emptying of the different anatomical levels of the lungs. This physiological pattern has traditionally been mimicked with IPPV. The technique of inversed I : E ratios was first reported in infants some 15 years ago (Reynolds 1971). Although a whole spectrum of inversed ratios has been investigated, oxygen delivery is optimal with ratios between 1.1 : 1 and 1.7 : 1 (Cole et al. 1984).

Our own investigations of three different I : E ratios demonstrate the effect on alveolar gas mixing efficiency and FRC using the nitrogen washout method described above. Twelve patients were ventilated with ratios of 1 : 2, 1 : 1 and 2 : 1. Six of these patients were ventilated postoperatively and had no known respiratory impairment whereas the six others were classified as patients in acute respiratory failure (ARF: $D(A-a)O_2 > 300$ mmHg). Alveolar gas mixing efficiency in the control group was found to be 67.5%, 70.5% and 73.6% and in the ARF group 48.5%, 51.9% and 54.0% with I : E ratios of 1 : 2, 1 : 1 and 2 : 1 respectively (Fig. 9.5). FRC rose in the first group from a mean value of 1970 ml to 2159 ml and 2341 ml, and in the ARF group from 2037 ml to 2177 ml and 2308 ml with increasing inspiratory times (Fig. 9.6). These results suggest that longer inspiratory times or breath-holding allow more time for diffusive mixing within the alveoli. The lower values for the ARF group show there is regional inhomogeneity of gas distribution and diffusive mixing, probably

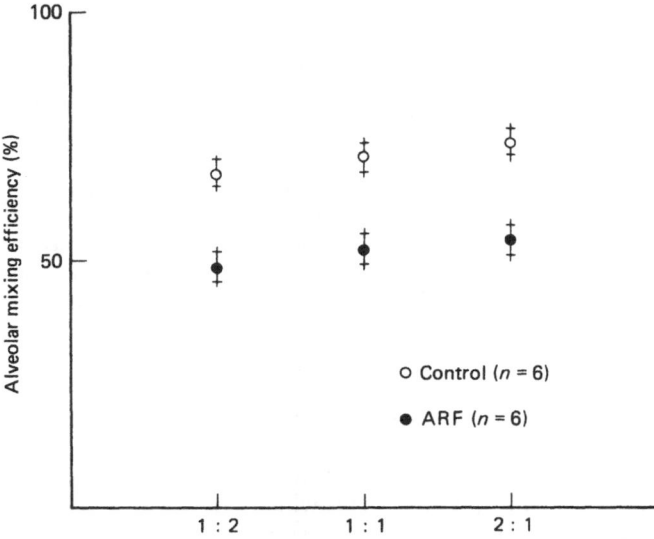

Fig. 9.5. Alveolar gas mixing efficiency for three different I : E ratios (1 : 2, 1 : 1 and 2 : 1) in patients with and without ARF.

due to alveolar collapse. The increase in FRC demonstrates an inadvertent PEEP effect during inversed ratio ventilation that occurs because of the shortened expiratory time. Since expiration during mechanical ventilation is passive the elastic recoil of the lungs determines expiration and leads to a "dynamic PEEP situation" when less time is allowed for expiration.

As shown by model analysis (Bowes et al. 1982) the amount and pattern of inspiratory flow can affect the functional dead space of the conducting airways, FRC, diffusive mixing and arterial oxygenation. With higher inspiratory flows the station-

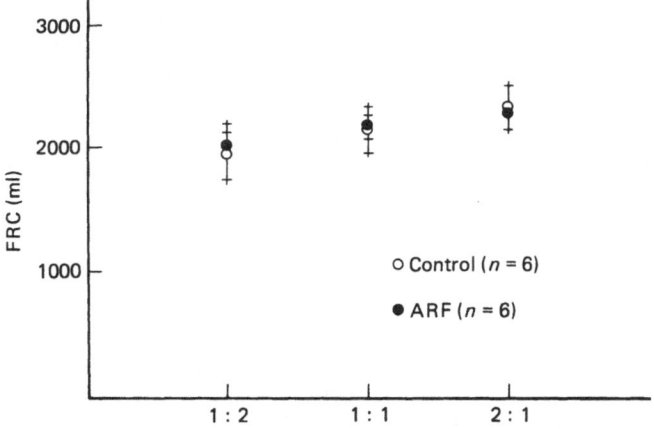

Fig. 9.6. FRC measurements in patients with and without ARF at I : E ratios of 1 : 2, 1 : 1 and 2 : 1.

ary interface described above moves further down the lungs during inspiration and expands the upper airways. If inspiratory time is increased this allows the interface to move back up the airways thus reducing anatomical dead space. The pattern of ventilation also has an impact on the dead space. When the rate of change of flow is higher, for instance a high flow at end-inspiration (increasing flow ramp), the interface will be established further down the lung. A high flow at the beginning of expiration as in inversed ratio ventilation results in the interface rapidly passing through the regions of the lungs where gas transport by molecular diffusion is favourable, hence enhancing diffusive mixing. Felton et al. (1984) found in patients after cardiopulmonary bypass surgery that when inspiratory flow was doubled arterial PO_2 increased and FRC fell. These changes were not due to alterations in ventilation inhomogeneity and they attributed them to differences in cardiac output or distribution of perfusion. Prolonged inspiratory times and high inspiratory flows, although having opposite effects on the FRC and dead space, appear to have the same effect on oxygenation. Advantages of inversed I : E ratios are the increase in FRC, and improvement in oxygenation and carbon dioxide elimination. Disadvantages may include depression of cardiac output and pulmonary barotrauma if mean airway pressure rises significantly. The same holds true for high inspiratory flows which can cause high airway pressures. At present inversed ratios or high flows should only be introduced in increments and oxygen delivery should be monitored. More work needs to be done on the best pattern of mandatory ventilation.

The Role of Spontaneous Breathing

The ultimate goal of respiratory support is the restitution of spontaneous breathing. It is therefore imperative to reduce the time of mandatory ventilation with its intrinsic pitfalls such as paralysis, sedation and the frequently mentioned "fighting the ventilator". The best ventilation obviously does not allow the ventilator to interfere with the patient's demands for adequate alveolar ventilation and oxygenation, as well as using the lowest possible effort to achieve this.

IMV (intermittent mandatory ventilation) has been advocated over the last decade to give the patient a minimum mandatory ventilation in order to maintain a spontaneous breathing pattern (Downs 1983). At the beginning of this chapter it was pointed out that during spontaneous breathing the pressure difference for inspiration and expiration is created from the alveolar level, where it most matters, and not, as in mechanical ventilation, from the mouth, where transmission of the pressure gradient becomes uncontrollable because of the existence of regions of lower resistance to air flow within the lungs. A step towards best ventilation is the maintenance of the spontaneous effort throughout the respiratory support by minimizing the work of breathing. The elastic work of breathing is proportional to the change in transpulmonary pressure and the tidal volume which is generated by the pressure gradient, as mentioned earlier. We have been led to believe that a sufficient decrease in lung compliance will increase the work of breathing and necessitate intubation and mechanical ventilation. But if FRC can be increased by the application of CPAP and sufficiently high inspiratory flows in excess of 150 l min^{-1} can be provided for inspiratory peak flows, a near-normal work of breathing can often be restored. Most of the so-called state-of-the-art ventilators are equipped with demand valves which deliver flows up to 120 l min^{-1}. If a patient requires peak flows in excess of these when inspiring rapidly, this will automatically lead to a transpulmonary pressure drop. Since the inspiratory and expiratory pressure difference is the determinant for the work of breathing for any given tidal volume, inspiratory flow rates should cater for the patient's demands. Clearly, the best system for IMV and CPAP consists of a continuous flow of gas and no flow resistance whatsoever.

One way of achieving this aim is the use of a continuous flow generator. The continuous flow can be in the region of 25–40 l min^{-1} if a reservoir bag of considerable size (> 10 l) is integrated into the inspiratory limb of the system (Bshouty et al. 1986). In order to minimize the inspiratory contribution to the work of breathing the reservoir bag should be positioned between a hinged and spring-loaded clapper-board to provide inspiratory support. By altering the strength of the springs the support can be changed according to the patient's requirements (Fig. 9.7). The gas flow should ideally be made available to the patient via a T-piece in order to avoid pressure created

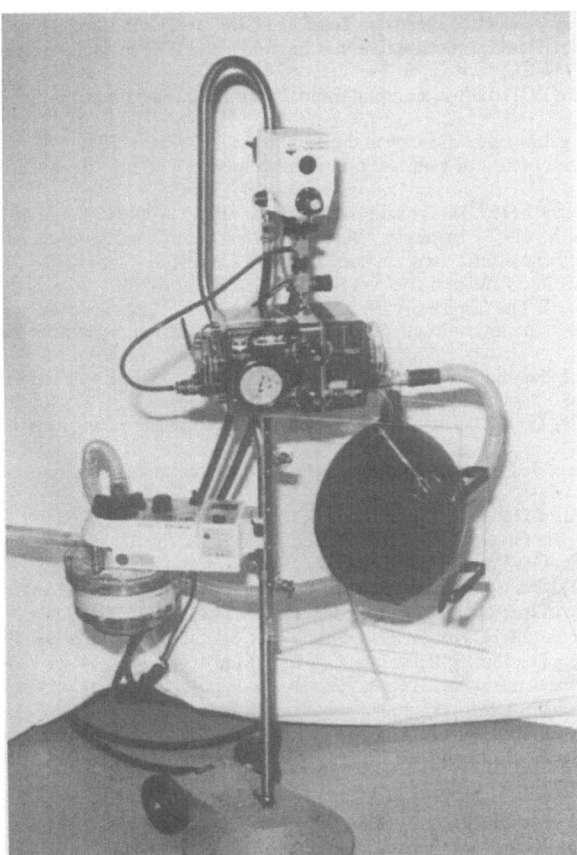

Fig. 9.7.   Continuous-flow CPAP circuit based on a Bird ventilator as high-flow generator, with 10 litre reservoir bag between hinged and spring-loaded clapper-boards.

by resistance, especially at the expiratory valve. To prevent mechanical failure of the exhalation valve CPAP should be generated by an underwater pipe. This provides the most accurate level of CPAP and the lowest risk of hyperinflation of the lungs.

In conclusion the optimum respiratory support may be defined as the CPAP system that will minimize airway pressure changes and hence the elastic work of breathing and pulmonary venous admixture, without causing adverse haemodynamic effects.

References

Artigas A, Carlet J, Chastang C, Le Gall JR, Cox P (1985) Protocol of adult respiratory distress syndrome study: clinical predictors, prognostic factors and outcome. European Society of Intensive Care Medicine

Bowes C, Cumming G, Horsfield K, Loughhead J, Preston S (1982) Gas mixing in a model of the pulmonary acinus with asymmetrical alveolar ducts. J Appl Physiol 52:624–633

Bshouty ZH, Roeseler J, Reynaert MS, Rodenstein D (1986) The importance of the balloon reservoir volume of a CPAP system in reducing the work of breathing. Intensive Care Med 12:153–156

Cole AGH, Weller SF, Sykes MK (1984) Inversed ratio ventilation compared with PEEP in adult respiratory failure. Intensive Care Med 10:227–232

Cumming G, Horsfield K, Preston SB (1971) Diffusion equilibrium in the lungs examined by nodal analysis. Respir Physiol 12:329–345

Cumming G, Guyatt A (1982) Alveolar gas mixing efficiency in the human lung. Clin Sci 62:541–547

Downs JB (1983) Ventilatory pattern and modes of ventilation in acute respiratory failure. Resp Care 28:586–591

Engel LA, Wood LDH, Utz G, MackCern PT (1973) Gas mixing during inspiration. J Appl Physiol 35:18

Felton CR, Montenegro HD, Saidel GM (1984) Inspiratory flow effects on mechanically ventilated patients: lung volume, inhomogeneity and arterial oxygenation. Intensive Care Med 10:281–286

Fischer AB (1980) Oxygen therapy: side effects and toxicity. Am Rev Respir Dis 122:61–69

Fowler WS (1948) Lung function studies. II. The respiratory dead space. Am J Physiol 154:405–461

Hillman DR (1986) Physiological aspects of intermittent positive pressure ventilation. Anaesthesiol Intensive Care 14:226–235

Ibanez J, Rawich JM, Moris SG (1983) Measurement of functional residual capacity during mechanical ventilation by simultaneous exchange of two insoluble gases. Anaesthesiol 54:413–417

Kox W, Langley F, Horsfield K, Cumming G (1982) The effect of atropine on alveolar gas mixing in man. Clin Sci 62:549–551

Krogh A, Lindhard J (1917) The volume of dead space in breathing and the mixing of gases in the lungs of man. J Physiol (Lond) 51:59–90

Lanczos C (1957) Applied analysis. Prentice-Hall mathematics series. Pitman, London

Loewy A (1894) Ueber die Bestimmung der Groesse des "Schaedlichen Luftraumes" im Thorax und der alveolaeren Sauerstoffspannung. Arch Ges Physiol 58:416–427

Ozanne GM, Zinn SE, Fairley HB (1981) Measurement of functional residual capacity during mechanical ventilation by simultaneous exchange of two insoluble gases. Anaesthesiol 54:413–417

Paloski WH, Newell JC, Gisser DG, Stratton HH, Annest SJ, Gottlieb ME, Shah DM (1981) A system to measure functional residual capacity in critically ill patients. Crit Care Med 4:342–346

Reynolds EOR (1971) Effect of alterations in mechanical ventilation settings on pulmonary gas exchange in hyaline membrane disease. Arch Dis Childh 46:152–159

Severinghaus JW (1966) Blood gas calculator. J Appl Physiol 4:102

Suter PM, Fairley HB, Isenberg MD (1975) Optimum end-expiratory airway pressure in patients with acute pulmonary failure. N Engl J Med 292:284–289

Wagner PD, Saltzman HA, West JB (1974) Measurement of continuous distributions of ventilation–perfusion ratios: theory. J Appl Physiol 36:588–599

West J (1979) Respiratory physiology. Williams and Wilkins, Baltimore

West J (1982) Ventilation–perfusion relationship. In: Scurr C, Feldman S (eds) Anaesthesia, scientific foundations. Heinemann, London

Chapter 10

The Arterial–Alveolar Nitrogen Difference for the Assessment of Ventilation–Perfusion Mismatch

P. Radermacher and K. J. Falke

Impaired pulmonary gas exchange leading to arterial hypoxaemia is one of the most common problems in critically ill patients. This arterial hypoxaemia and, hence, an increased alveolar–arterial oxygen partial pressure difference $(D(A–a)O_2)$—at a given fraction of inspired oxygen (FiO_2) and a stable cardiac output—are mainly attributed to ventilation–perfusion mismatch (low V_A/Q areas) and direct intrapulmonary right-to-left shunting. The discrimination of these two components has been of interest as their causes and treatment may be different (Bendixen 1964; Pesenti et al. 1983). Berggren (1942) tried to separate the two effects by calculating the venous admixture (Q_{VA}/Q_T) with and without nitrogen present in the inspired gas. In practice this method yielded contradictory results, both enhanced and decreased Q_{VA}/Q_T values being found (Quan et al. 1980; Shapiro et al. 1980; Suter et al. 1985), and evidence for absorption collapse of alveoli being observed (Dantzker et al. 1975; Wagner et al. 1974a). Moreover, increasing the oxygen tension in mixed venous blood by pure oxygen breathing may increase the right-to-left shunt (Bishop and Cheney 1983) by reversing hypoxic pulmonary vasoconstriction in unventilated lung areas (Domino et al. 1983).

The Nitrogen Gradient

Canfield and Rahn (1957) studied the arterial and alveolar nitrogen partial pressures (PN_2) to assess the amount of ventilation–perfusion variation in the lungs: whenever the exchange rates of oxygen and carbon dioxide are not equal in a single gas exchange unit the nitrogen fraction is altered from that in the inspired air. Thus the lung acts as a population of individual gas exchange units with varying nitrogen partial pressures, the net exchange rate being zero. In alveoli with low V_A/Q ratios the PN_2 rises almost as fast as the PO_2 declines, as the PCO_2 can increase by only a few millimetres of mercury to the mixed venous value. As these alveoli with high PN_2 values contribute relatively more to the perfusion than the alveoli with high V_A/Q (and hence lower PN_2) an arterial–alveolar nitrogen partial pressure difference $(D(A–a)N_2)$ comes into existence (Canfield and Rahn 1957; Rahn and Farhi 1964).

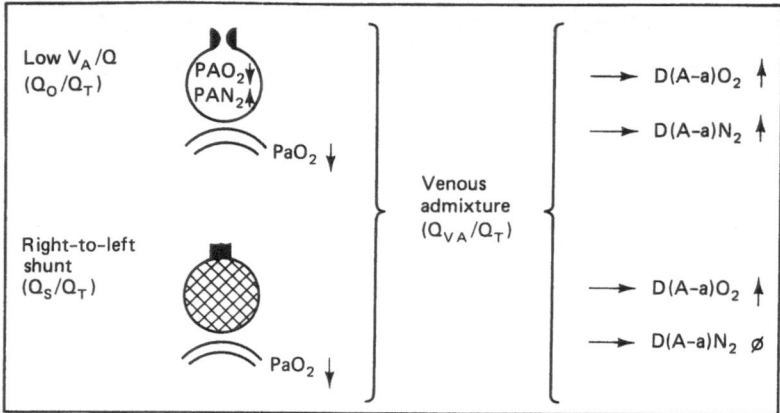

Fig. 10.1.   Effect of low V_A/Q areas and true right-to-left shunting on the venous admixture (Q_{VA}/Q_T), the alveolar–arterial difference for PO_2 $(D(A-a)O_2)$ and the arterial–alveolar difference for PN_2 $(D(A-a)N_2)$.

As there is neither production nor consumption of nitrogen in the tissues, the arterial and the mixed venous PN_2 must be equal (Canfield and Rahn 1957; Rahn and Farhi 1964). Therefore the $D(A-a)N_2$ is not influenced by direct right-to-left shunting and reflects only low V_A/Q areas (Canfield and Rahn 1957; Rahn and Farhi 1964) (Fig. 10.1).

Using a manometric technique (Klocke and Rahn 1961a) with a Van Slyke apparatus (Van Slyke and Neill 1924) Klocke and Rahn (1961b) determined the $D(A-a)N_2$ in normal and emphysematous subjects after measuring the urine PN_2. A modification of the technique allowed its application to samples of whole blood (Klocke and Rahn 1960). In normal subjects the average $D(A-a)N_2$ was 3.4 mmHg, whereas in the emphysematous patients it was 22.2 mmHg. Farhi and coworkers (1963) found a normal value for the $D(A-a)N_2$ of 9.4 mmHg by applying a combined technique of vacuum extraction and gas chromatography analysis to the measurement of the arterial PN_2.

Lenfant (1963, 1965) studied the effect of increasing the FiO_2 on the $D(A-a)N_2$ in normal and diseased subjects. He showed that, as does the $D(A-a)O_2$, the $D(A-a)N_2$ depends upon the FiO_2. When the FiO_2 is raised both the $D(A-a)O_2$ and the $D(A-a)N_2$ rise concomitantly in lung compartments with a V_A/Q greater than 0.01; because of the non-linear shape of the haemoglobin oxygen dissociation curve the $D(A-a)O_2$ rises faster than the $D(A-a)N_2$ in regions with V_A/Q less than 0.01 (Lenfant 1965). Consequently the $D(A-a)N_2$ can never be higher than the $D(A-a)O_2$.

The "Nitrogen Shunt" Concept

In order to compare the data from the nitrogen partial pressure analyses with the venous admixture values Markello and coworkers (1972) used the concept of the "nitrogen shunt" in critically ill patients, particularly trauma patients (Markello et al. 1974). Lenfant (1965) referred to the nitrogen shunt as "the dependent flow to alveoli with indeterminably low V_A/Q". Assuming a two-compartment lung model, one compartment having a V_A/Q ratio of approximately 1, the other being perfused and open but unventilated (i.e. $V_A/Q = 0$), the fractional blood flow to the unventilated compartment (nitrogen shunt, Q_O/Q_T) can be calculated by:

$$Q_O/Q_T = \frac{D(A–a)N_2}{P_{baro} – PH_2O – P\bar{v}O_2 – P\bar{v}CO_2 – PAN_2}$$

where PAN_2 is the alveolar nitrogen partial pressure, $P\bar{v}O_2$ is the mixed venous oxygen partial pressure and $P\bar{v}CO_2$ the mixed venous carbon dioxide partial pressure.

The above-mentioned normal values of 3 and 9 mmHg for the $D(A–a)N_2$ yield Q_O/Q_T values of 6% and 16% respectively. In contrast to the measurement of the $D(A–a)N_2$ the calculation of the Q_O/Q_T value is not dependent upon the FiO_2.

In 24 of the 30 patients studied Markello et al. obtained $D(A–a)N_2$ data exceeding the normal value, which thus indicated ventilation–perfusion mismatch. Applying the mixing equation for the two-compartment lung model Q_O/Q_T values up to 46% were computed. In some patients there were considerable differences between the values of Q_O/Q_T and Q_{VA}/Q_T. This was explained by the fact that "both models misrepresent reality" (Markello et al. 1972). The compartment with $V_A/Q = 0$ in reality being a lung region with very low V_A/Q ratios ($0.01 < V_A/Q < 0.1$), the real alveolar PO_2 in this region is slightly higher than the $P\bar{v}O_2$—especially when the FiO_2 reaches 0.8—which causes incomplete haemoglobin saturation and reduces the Q_{VA}/Q_T value. Because of the steep slope of the haemoglobin dissociation curve these minor differences in alveolar PO_2 induce considerable differences in blood oxygen content while the difference between the PN_2 in this region and the mixed alveolar gas (and hence the $D(A–a)N_2$) is only slightly influenced. Consequently the value of Q_O/Q_T may greatly exceed the value of Q_{VA}/Q_T (Markello et al. 1972).

Recently we measured (unpubl. data) the $D(A–a)N_2$ and the sulphur hexafluoride (SF_6) retention in order to evaluate the contribution of low V_A/Q areas and true right-to-left shunting (Q_S/Q_T) to the enhanced $D(A–a)O_2$ and Q_{VA}/Q_T in 15 critically ill patients who were either mechanically ventilated or breathing spontaneously with continuous positive airway pressure (CPAP). The impairment of pulmonary gas exchange was due either to one of those phenomena or a combination of both.

We did not find any correlation between Q_O/Q_T and the difference between Q_{VA}/Q_T and Q_S/Q_T because these values are computed according to different lung models. Furthermore, the determinations of $Q_S/Q_T(SF_6)$ and $Q_O/Q_T(N_2)$ are not completely distinct from each other. Because of its finite—though very low—solubility SF_6 provides an index not only of true right-to-left shunting but of all lung compartments having a $V_A/Q < 0.005$. Compartments with $0 < V_A/Q < 0.005$, however, also increase the $D(A–a)N_2$ and hence the Q_O/Q_T value.

Conclusion

Because of varying theoretical assumptions the measurement of the $D(A–a)N_2$ and the calculation of the nitrogen shunt Q_O/Q_T cannot yield an exact quantification of the fractional blood flow to all underventilated lung areas as is provided by the multiple inert gas elimination technique of Wagner et al. (1974a). Nevertheless, they allow an improved estimation of the contribution of ventilation–perfusion mismatch to arterial hypoxaemia when compared with calculation of the physiological oxygen shunt. Being less complicated than the multiple inert gas technique the method may be applicable to clinical research.

This work was supported by the Deutsche Forschungsgemeinschaft, AZ.:FA139/1-1.

References

Bendixen H (1964) Atelectasis and shunting. Anesthesiology 25:595–596

Berggren S (1942) The oxygen deficit of arterial blood caused by non-ventilating parts of the lung. Acta Physiol Scand [Suppl] 11:1–92

Bishop M, Cheney F (1983) Effects of pulmonary blood flow and mixed venous O_2 tension on gas exchange in dogs. Anesthesiology 58:130–135

Canfield R, Rahn H (1957) Arterial–alveolar N_2 gas pressure difference due to ventilation perfusion variation. J Appl Physiol 10:165–172

Dantzker D, Wagner P, West J (1975) Instability of lung units with low V_A/Q ratios during O_2 breathing. J Appl Physiol 38:886–895

Domino K, Wetstein L, Glasser S et al. (1983) Influence of mixed venous oxygen tension ($P\bar{v}O_2$) on blood flow to atelectatic lung. Anesthesiology 59:428–434

Farhi L, Edwards A, Homma T (1963) Determination of dissolved N_2 in blood by gas chromatography and (a–A) N_2 difference. J Appl Physiol 18:97–106

Klocke F, Rahn H (1960) The arterial–alveolar inert gas pressure difference. In: Studies in Pulmonary Physiology. Wright Air Development Center Technical Report 60–1, pp 66–90

Klocke F, Rahn H (1961a) A method for determining inert gas ("N_2") solubility in urine. J Clin Invest 40:279–285

Klocke F, Rahn H (1961b) The arterial–alveolar inert gas ("N_2") difference in normal and emphysematous subjects by the analysis of urine. J Clin Invest 40:286–294

Lenfant C (1963) Measurement of ventilation/perfusion distribution with alveolar–arterial differences. J Appl Physiol 18:1090–1094

Lenfant C (1965) Effect of high FiO_2 on measurement of ventilation perfusion distribution in man at sea level. Ann NY Acad Sci 121:797–808

Markello R, Winter P, Olszowka A (1972) Assessment of ventilation–perfusion inequalities by arterial–alveolar nitrogen differences in intensive care patients. Anesthesiology 37:4–15

Markello R, Schuder R, Border J (1974) Arterio-alveolar N_2 differences documenting ventilation–perfusion mismatching following trauma. J Trauma 14:423–426

Pesenti A, Riboni A, Marcolin R, Gattinoni I (1983) Venous admixture (Q_{VA}/Q_T) and true shunt (Q_S/Q_T) in ARF patients: effects of PEEP at constant F_IO_2. Intensive Care Med 9:307–311

Quan S, Kronberg G, Schlobohm R, Feeley T, Don H, Lister G (1980) Changes in venous admixture with alterations of inspired oxygen concentration. Anesthesiology 52:477–482

Rahn H, Farhi L (1964) Ventilation, perfusion and gas exchange—the V_A/Q concept. In: Handbook of physiology, section 3, Respiration I. American Physiological Society, Washington DC, pp 735–766

Shapiro B, Cane R, Harrison R, Steiner M (1980) Changes in intrapulmonary shunt with administration of 100% oxygen. Chest 77:138–141

Suter P, Fairley B, Schlobom R (1975) Shunt, lung volume and perfusion during short periods of ventilation with oxygen. Anesthesiology 43:617–627

Van Slyke D, Neill J (1924) The determination of gases in the blood and other solutions by vacuum extraction and manometric measurement. J Biol Chem 61:523–573

Wagner P, Saltzmann H, West J (1974a) Measurement of continuous distributions of ventilation–perfusion ratios. J Appl Physiol 36:588–599

Wagner P, Laravuso R, Uhl R, West J (1974b) Continuous distributions of ventilation–perfusion ratios in normal subjects breathing air and 100% O_2. J Clin Invest 54:54–68

Chapter 11

High-frequency Ventilation: A Step Towards "Compliance-Independent" Ventilation

P. P. Lunkenheimer, W. F. Whimster, N. Stroh, J. Theissen, G. Frieling and H. Van Aken

From its very beginning high-frequency ventilation has required a thorough reinvestigation of the basic mechanisms of gas exchange and alveolar ventilation during spontaneous or any kind of assisted ventilation.

In 1955 Emerson patented oscillatory pressure excitation as a form of mechanical ventilation (Emerson 1959), although he did not demonstrate that this method was an efficient means of ventilation. In 1967 Sanders used a jet method to ventilate patients during bronchoscopy. A highly accelerated air bolus was injected intermittently into the lumen of the endoscope. The method was widely used but popularized by the many Scandinavian investigators and has now become the most common variation of high-frequency ventilation. Öberg and Sjöstrand introduced high-frequency jet ventilation in 1968. Their aim was to increase the ventilatory rate to 150 min^{-1} in order to reduce pulsations in the carotid artery during neuro-electrophysiological investigations. Tidal volume was maintained more or less within the physiological range resulting in a very high minute volume. Gas injection was governed by electromagnetic valves. The method was soon introduced into clinical practice, namely as a form of ventilatory assist during bronchoscopy.

Guided by these promising results with high-frequency jet ventilation, anaesthetists tended to increase the ventilatory frequency in a modification of assisted ventilation. However, these tentative steps in clinical experimentation produced variable results which were difficult to interpret (Baum et al. 1980).

High-frequency oscillation originally consisted of low tidal volume oscillations induced to produce intrathoracic pressure oscillations (Lunkenheimer et al. 1972, 1978). We used the method as a form of cardiodiagnostic monitoring to assess transmyocardial pressure transmission in order to measure myocardial stiffness. The effect this method had on lung ventilation was unexpected.

High-frequency oscillation was first used in paediatric intensive care by Bohn et al. in 1979 (Bohn et al. 1980). The method is able to normalize the carbon dioxide and oxygen partial pressures in children (as well as in small dogs weighing less than 15 kg). However, while gas exchange is adequate in some adults, in others it is not or the method works for a few hours only.

The clinical and experimental experience of an inhomogeneous response to any kind of high-frequency ventilation has led to some investigators increasing the tidal

volume. However, this only produces a more and more accelerated bulk flow characterized by an increasing selectivity of the bundled jet which then predominantly ven-
tilates the dependent, i.e. the lower lung, compartments. Because the beneficial
effects have not been uniform or consistent, high-frequency ventilation is not yet
accepted as a standard method of ventilation in clinical practice. We therefore set out
to investigate whether the controversial and irreproducible results of high-frequency
ventilation were due to the characteristics of the exciting systems, variations in the
anatomy of the respiratory system, or variations in the dynamic responses of individual respiratory systems to this form of excitation.

Methods

Exciting Systems

We used the Acutronic ventilator which works at a frequency range of 150–600 min^{-1}
and a mean gas flow of 5–36 l min $^{-1}$, depending on the driving pressure, the exciting
frequency (at 2 Hz flow is 36 l min^{-1}, at 10 Hz it is 20 l min^{-1}) and the inspiratory to
expiratory time (I:E) ratio. High-frequency oscillation was induced by an eccentric
piston pump at a tidal volume of 25 ml and a frequency between 5 and 63 Hz.

In order to eliminate the bundled jet characteristics an endotracheal rotating valve
tube consisting of a 9 mm external diameter tube was developed. At the end of this
tube a propeller-like spray nozzle was fitted which conveys the air in a widespread
bulk form to the central bronchi (Fig. 11.1). A second port facing backwards towards
the tube's lumen was opened intermittently in a rhythm interposed between two
orthograde injections, using the same rotating disc driven by a flexible axle. Up to 76
disc rotations per second were used, giving up to 76 jets forwards and backwards per
second. The mean orthograde and backwards flow reached up to 60 l min^{-1} depending on the driving pressure. In the animal experiments the rotation frequency was
varied between 20 and 76 rotations per second (3–5 times per minute).

Models

The model lungs were obtained post mortem from three different species: dogs, pigs
and calves. In these lungs the trachea was cannulated. They were inflated with air
until all visible segments were aerated. Air escaped via the pulmonary veins. A continuous flow of air was maintained at the distending pressure for 48 hours, by which
time the lungs were dry and the weight was constant at about 15% of the wet weight.

Intrapulmonary Pressure

Some of these dried lungs were used to measure parenchymal air pressure. In order
to obtain access to the subpleural parenchyma the wide ends of several Statham
domes were fixed to the exposed lung surface (pleura removed) which was
then coated with silicone rubber. The air pressure in the underlying parenchyma was
measured at the site of each dome.

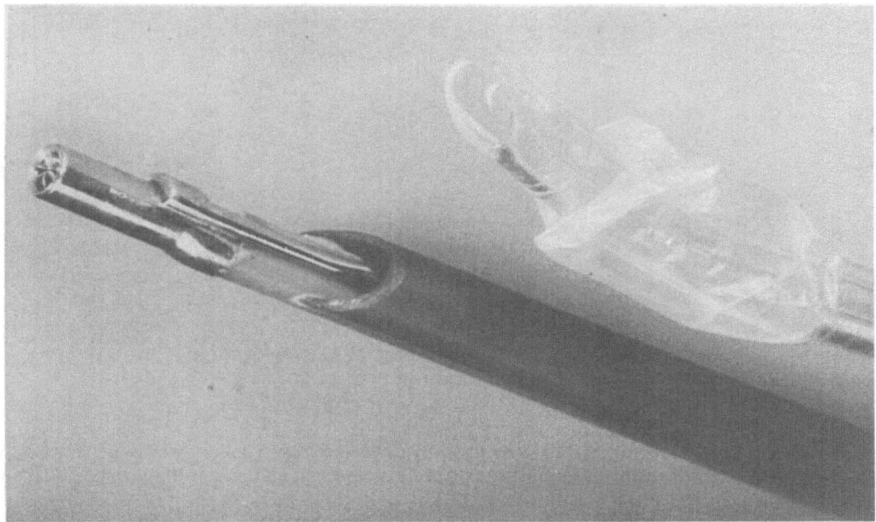

Fig. 11.1.   The rotating disc valve catheter, which is integrated into a conventional endotracheal tube. A distal spray nozzle conveys the air in a widespread bulk to the central bronchi. A second port facing backwards towards the lumen of the tube is used to create a sucking phase, interposed between two orthograde insufflations.

Bronchial Aerodynamics

Other dried respiratory systems were used to study bronchial aerodynamics. In some, the parenchyma was removed from the bronchial tree with forceps, while in others the parenchyma was blown off with a sandblaster. The exposed bronchi were separated into six "lobes" by covering the ends with colostomy bags (Fig. 11.2). The pressure in the bags produced by ventilation through the trachea was measured.

Transparenchymal Carbon Dioxide Counterflow

Transparenchymal carbon dioxide counterflow was studied by fixing a cannula from a carbon dioxide supply to one opening of one Statham dome on the pleural surface. The other port of that dome was open to atmosphere (Fig. 11.3). The parenchyma exposed within the dome was then equilibrated with carbon dioxide (flow 200 ml min^{-1}) at atmospheric pressure. The concentration of carbon dioxide within the trachea was measured under three different conditions:

1. With the trachea open to the atmosphere with no inflow or ventilation
2. With 2 l min^{-1} of air insufflated into the trachea
3. With high-frequency oscillation superimposed on condition 2.

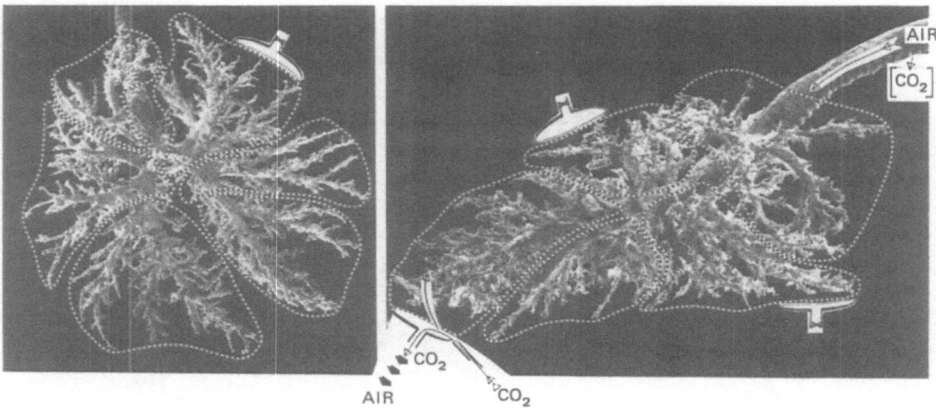

Fig. 11.2. The isolated bronchial system schematically divided into six "lobes" which are wrapped in transparent (colostomy) bags. Domes for flow and pressure measurements, gas insufflation and gas sampling are integrated in the wall of each sack.

Animal Experiments

In 7 dogs of either sex weighing 35–48 kg selective catheterization of three to five pulmonary veins, left atrium, left ventricle and of the central pulmonary artery was performed by left-sided thoracotomy. Samples for blood gases were taken during different high-frequency excitation patterns.

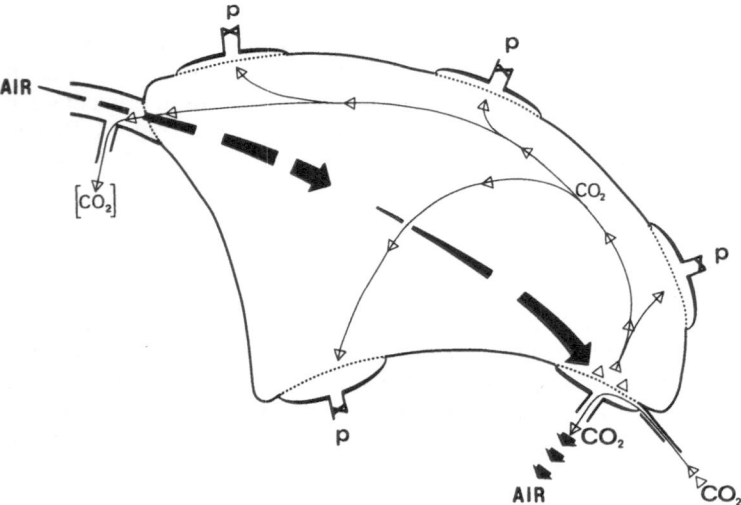

Fig. 11.3. Diagram of the dried lung model coated with silicone rubber and with domes for measurements sealed to the parenchyma after removal of the pleura. The procedure for the carbon dioxide counterflow experiments is shown schematically.

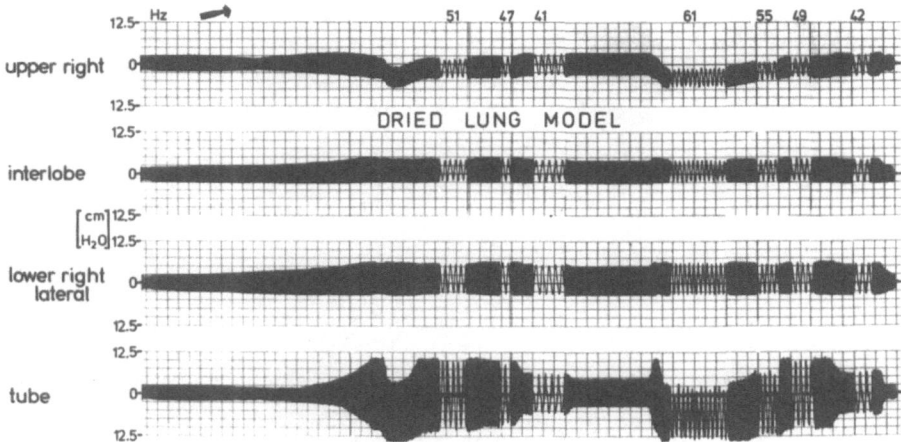

Fig. 11.4. Original recording of pressure changes in different compartments of the dried lung model during high-frequency oscillation excitation at rising frequencies.

Results

Model Experiments on Intrapulmonary Pressure

High-Frequency Oscillation

Fig. 11.4 shows a representative recording of the changes in pressure amplitude and mean pressure at four sites in one respiratory tract. The results reveal inhomogeneities in amplitudes as frequency increases and varying patterns of mean pressures ranging from below zero pressure, i.e. sucking, to positive pressures, i.e. airtrapping (Fig. 11.5). When the conditions were changed by altering the tube's position in the trachea, by retracting or rotating it, the pattern of pressure amplitudes and mean pressures changed less markedly than with changes in excitation frequency.

High-Frequency Jet Ventilation

With high-frequency jet ventilation using frequencies up to 10 Hz, no drop in mean pressure below zero was observed but there was a marked dependence on frequency, i.e. a decrease in pressure amplitude with frequency. This was because the jet ventilator used produced a lower maximum gas flow when the frequency was increased from 2 to 10 Hz, namely 36 and 20 l min^{-1} respectively. However, the patterns of pressure amplitudes and mean pressures were much less dependent on changes in frequency than they were with high-frequency oscillation, but were all the more dependent on the position of the tube in the trachea, i.e. on the direction of the jet.

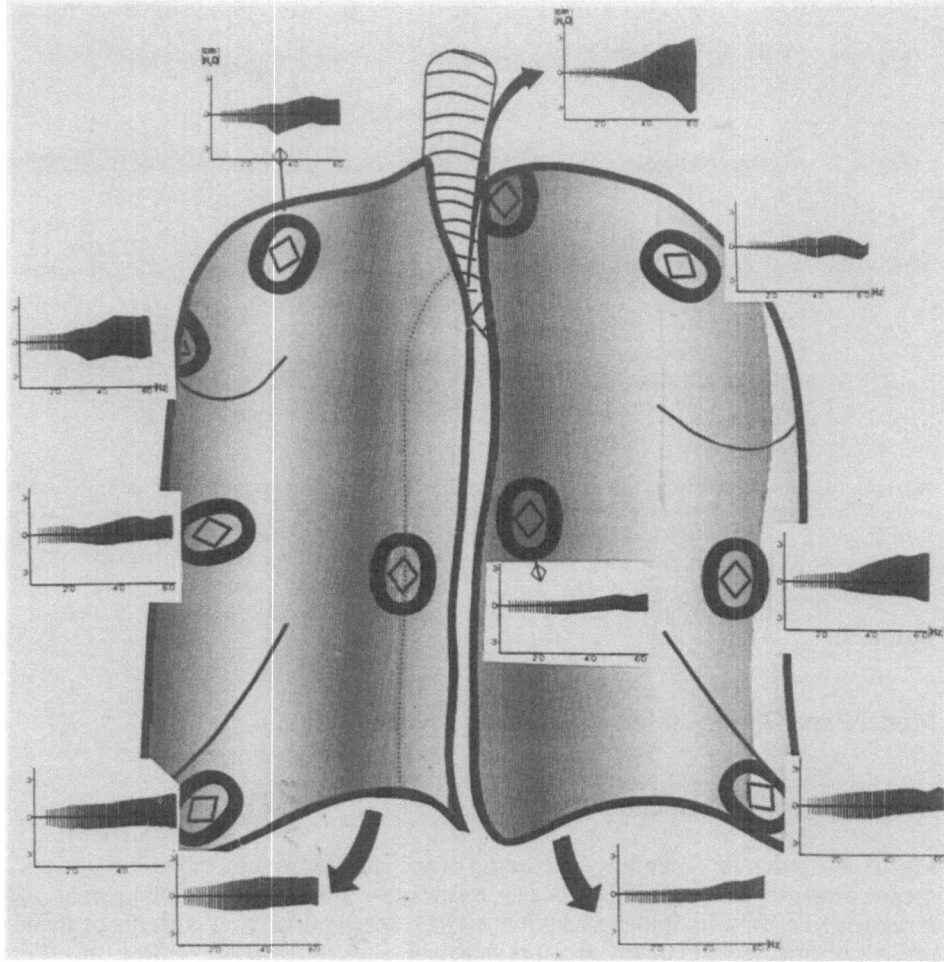

Fig. 11.5. Pressure pattern (mean and amplitudes) in different compartments of the dried lung model during high-frequency oscillation excitation at rising frequencies (3–60 Hz).

Endotracheal High-Frequency Ventilation

With high-frequency ventilation induced by the bidirectional rotating valve tube, the central pressure changes were transmitted well to the peripheral compartments (Fig. 11.6). The pressure impulses contained a mixture of frequencies comparable to those generated by a random signal generator. We were able to manipulate mean pressures freely above zero and below zero by varying the mean backwards flow volume. Frequency-dependent inhomogeneities in lobar ventilation were less marked than in high-frequency oscillation where the excitation pattern was sinusoidal. However, a certain dependence on the position of the tube was observed only when the rotating valve was almost touching the carina.

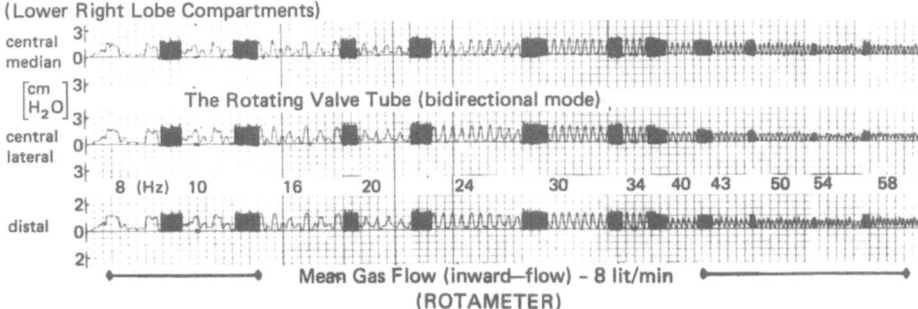

Fig. 11.6. Peripheral pressure recording at three measuring sites of the dried right lower lobe during high-frequency excitation by the rotating disk valve catheter.

Bronchial Aerodynamics

Results obtained from the isolated bronchial preparation with high-frequency oscillation, high-frequency jet ventilation and endotracheal high-frequency ventilation were comparable with those obtained from the parenchymal dome preparations, except that all the pressure amplitudes were up to 30% greater.

Transparenchymal Carbon Dioxide Counterflow (Fig. 11.3)

With the trachea open to the atmosphere, carbon dioxide was detectable within the trachea when the surface of the parenchyma under the dome was exposed to a carbon-dioxide-rich atmosphere. When air was insufflated into the trachea it was washed out of the parenchyma through the open duct of the dome to the atmosphere.

When high-frequency oscillation or endotracheal excitation was superimposed on the insufflated air, carbon dioxide was again detected in the trachea. Thus a counter-current of carbon dioxide was set up from the dome through the parenchyma to the trachea against the insufflating flow from the trachea to the dome. Pure high-frequency jet ventilation (no sucking phase) was not able to evoke the carbon dioxide counterflow.

Animal Experiments

Inhomogeneities in lobar ventilation (Fig. 11.7), oxygen uptake and carbon dioxide elimination differed from one measuring site to another, i.e. between different lobes. This was true for spontaneous and conventional mechanical ventilation, but it was particularly increased with high-frequency jet ventilation and oscillation. The ventilatory pattern changed essentially with the mean intrapulmonary pressure and with the exciting frequency, but the resulting changes were not consistent, i.e. not reproducible nor predictable. Endotracheal excitation by the rotating valve tube (at systematically varied frequencies between 20 and 76 Hz) induced a more homogeneous

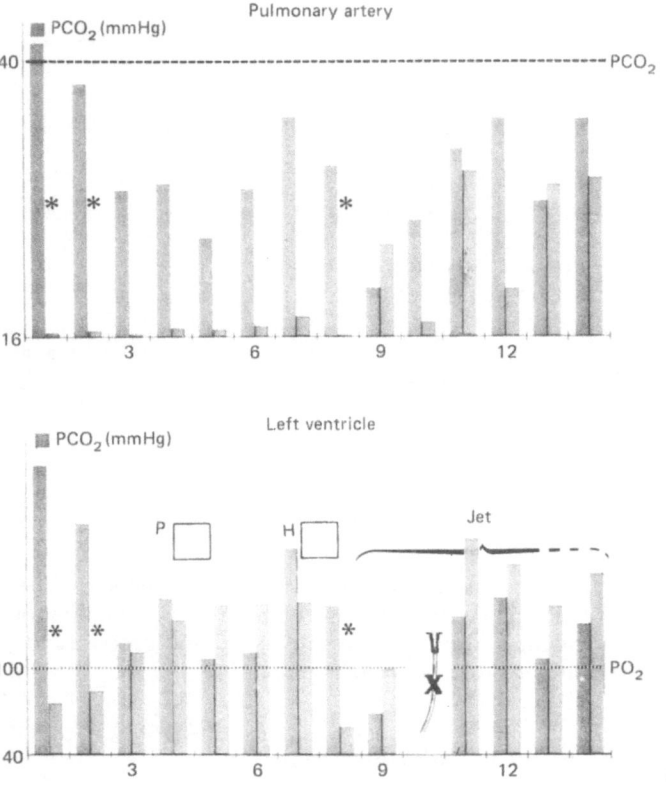

Fig. 11.7. Blood gas measurements (left column, PCO_2; right column, PO_2) in the pulmonary artery and left ventricle (**a**) and in the pulmonary veins of the left upper and middle and of the right lower lobe of the lung (**b**) during different types of spontaneous breathing (*) and mechanical ventilation (P, pulmomat; H, high-frequency oscillation excitation and high-frequency jet ventilation with an acutronic ventilator). During jet ventilation the position of the tube was changed by pulling it back and pushing it forward and by turning the curvature to the right and to the left corner of the mouth.

pattern of lobar ventilation. Nevertheless, frequency and mean pressure dependency of lobar and global gas exchanges was still present.

Discussion

Our experiments confirm the inhomogeneities in the pattern of lobar ventilation produced by high-frequency ventilation. Two main reasons for these inhomogeneities emerge: (a) frequency dependence, particularly with frequencies above 15 Hz, and (b) position of the tube in the trachea, which although apparent with high-frequency oscillation seems to be critical with high-frequency jet ventilation because of the

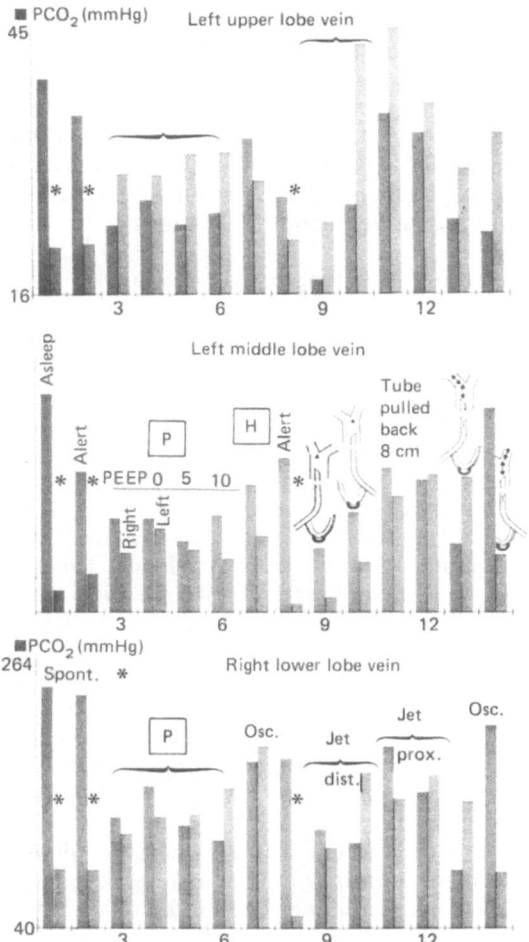

Fig. 11.7b

directional selectivity of the accelerated gas flow. Endotracheal high-frequency ventilation may produce a less narrow directional jet.

The potential beneficial effects of a biphasic pressure impulse are produced in the endotracheal excitation technique by alternating backward and forward flows. The non-sinusoidal impulse consists of a mixture of high and low frequencies. Taking advantage of the frequency dependency of lobar ventilation, endotracheal high-frequency ventilation aims to excite each lobe at its own natural frequency.

The same general effect might be expected with jet ventilation, which is also a form of impulse ventilation. However, this potential advantage of global excitation of the lung by the jet impulse is abolished by its marked directional selectivity. The accelerated bulk flow down to the lower lobes of both lungs is likely to dominate intrapulmonary gas re-partition. Resulting volume changes of the epidiaphragmatic lung compartments might secondarily act on the rest of the lung volume producing rhythmic compression and decompression. Inflation of the lower lobes could induce deflation of the upper lobes and vice versa. The accelerated jet down the long axis of the airways may induce Venturi effects which possibly suck on the upper lobe bronchi. The resulting internal contamination of the injected gas bolus with carbon dioxide from the upper lobes (*Pendelluft*) would explain the relatively low efficiency of high-frequency jet ventilation in the elimination of carbon dioxide. However, the method still has particular advantages when local lung tissue mobilization is required by alveolar hyperventilation of basal lung compartments. Thus, jet ventilation seems to be predominantly some kind of conventional ventilation characterized by compartmental selectivity.

The rigid bronchial system as studied using dried lung preparations is clearly not the same as the respiratory system examined in vivo. The in vitro preparations suggest that during oscillatory excitation of the lung the main source of inhomogeneities in alveolar ventilation is a frequency-dependent intrabronchial gas flow phenomenon. Inhomogeneities are also present in the isolated bronchial preparations, which means that the inhomogeneous mechanical properties of the lung parenchyma are not the only factors involved.

Inhomogeneities of ventilation produced by high-frequency ventilation can be used to advantage, as well as to explain the frequent clinical failure of the method. To be useful the inhomogeneities have to be manipulated systematically to induce intercompartmental gas flows as a means of gas mixing within the parenchyma. This calls for systematic changes in frequency and prevention of preferential ventilation resulting from the direction of the jet. Ventilation thus becomes less dependent upon local elasticities and tissue compliances. Alveolar ventilation is then determined by the pattern of frequency-dependent endobronchial resistances. For this purpose high frequencies exceeding 15 Hz, changed periodically and given maximum freedom from directional constraints, are required to homogenize alveolar ventilation. The endotracheal high-frequency excitation system is the system of choice for its combination of frequency characteristics, its biphasic pressure effects, and its reduced directional selectivity.

The development of suitable monitoring systems to detect and measure inhomogeneities in alveolar ventilation is essential in order to allow their systematic manipulation by changing the exciting frequency. This is the next step to be taken before high-frequency ventilation can be recommended for widespread clinical application.

References

Baum M, Benzer HF, Geyer M, Haider W, Mutz N (1980) Forcierte Diffusionsventilation (FDV). Grundlagen und Anwendung. Anaesthesist 29:586–591

Bohn DJ, Miyasaka K, Merchak BE, Thompson AK, Froese AB, Bryan AC (1980) Ventilation by high-frequency oscillation. J Appl Physiol 48:710–716

Emerson JH (1959) Apparatus for vibrating portions of a patient's airways. US Patent Office: Application 2 March 1955, Ser. No. 491,699; Patented 29 Dec. 1959, No. 2,918,917.

Lunkenheimer PP, Raffenbeul W, Keller H, Frank J, Dickhuth HH, Fuhrmann C (1972) Application of transtracheal pressure oscillations as a modification of "diffusion respiration". Br J Anaesthesiol 44:627

Lunkenheimer PP, Ising H, Frank J, Scharsich M, Dittrich H (1978) Enhancement of CO_2-elimination by intrapulmonary high frequency pressure alternation during "apneic oxygenation". In: Oxygen transport to tissue, Vol 3. Plenum Press, New York, pp 599–603

Öberg PA, Sjöstrand U (1969) Studies of blood-pressure regulation: II. Acta Physiol Scand 75:287–300

Sanders FD (1967) Two ventilation attachments for bronchoscopes. Delaware Med J 39:170–175

Chapter 12

Extracorporeal Support in Acute Respiratory Failure

L. Gattinoni, A. Pesenti, R. Marcolin, D. Mascheroni, R. Fumagalli,
A. Riboni, F. Rossi, F. Scarani, L. Avalli and A. Giuffrida

Following the introduction of the membrane lung into clinical practice it was appealing to use this device as a substitute for alveolar capillary function in severe acute respiratory failure (ARF). Conventional treatment by positive pressure ventilation does not, in fact, substitute for respiratory function, but maximally exploits the residual gas-exchanging regions of the diseased lungs. Hill et al. (1972) reported the first successful clinical application of the membrane lung in the treatment of ARF but, from 1966 to 1975, only a 10% survival rate was obtained in 233 cases treated worldwide (Gille and Bagniewski 1976).

In 1974–1977 the National Heart and Lung Institutes (Bethesda, USA) funded a multi-centre trial (NHLI 1974) to compare extracorporeal membrane lung oxygenation (ECMO) and continuous positive pressure ventilation (CPPV) with CPPV alone in ARF. The patients were selected according to gas exchange pattern under standardized ventilatory conditions applied for defined periods of time (2 hours, fast entry; 48 hours, slow entry criteria). These criteria were originally chosen to detect a population with a predicted mortality rate of about 50%. Surprisingly, the mortality rate in the CPPV group was 90%, and the same mortality was observed in the ECMO–CPPV group (Zapol et al. 1979).

The ECMO trial demonstrated that extracorporeal support could safely be performed for a prolonged period of time, but, unfortunately, it did not increase the survival rate. It seemed that substitution of respiratory function with the membrane lung and CPPV was not useful in ARF. However, it is possible that the potential of this form of support was not fully exploited during the ECMO trial.

In 1979 (Gattinoni et al. 1980) we introduced low-frequency positive pressure ventilation (LFPPV) with extracorporeal carbon dioxide removal ($ECCO_2R$) as a different form of extracorporeal support. In this chapter we shall describe the rationale of this technique, the technique itself and the clinical results obtained in 48 patients with severe ARF.

ARF Lung: The Pathological and Functional Model

Fig. 12.1a shows a typical example of ARF lung during mechanical ventilation, as imaged with a computerized tomographic (CT) scan. Fig. 12.1b shows the same lung following an increase in positive end-expiratory pressure (PEEP) from 5 to 15 cm H_2O while maintaining the total ventilation constant. We recently reported a series of ARF patients studied at different PEEP levels by CT scan (Gattinoni et al. 1986) and found, on average, the same morphological distribution shown in Fig. 12.1. The ARF lung appears to be made up of three different zones: one healthy (zone H), a second recruitable (zone R), and a third diseased (zone D). The existence of such zones may be inferred from other available data independently of CT imaging.

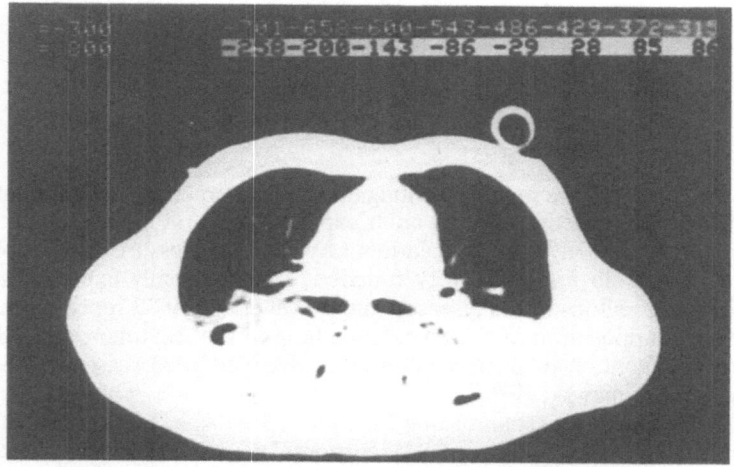

a

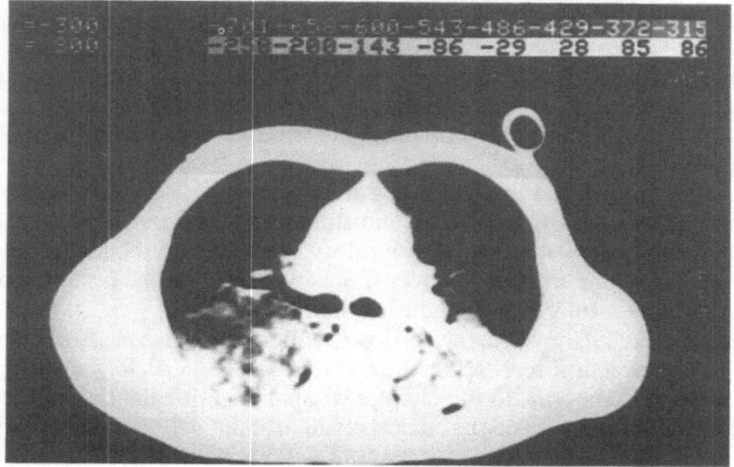

b

Fig. 12.1.   CT scan of ARF lung: **a** at a PEEP of 5 cm H_2O; **b** at a PEEP of 15 cm H_2O.

Zone H

Radiology, pathological samples and CT imaging document the presence of areas of lung in ARF which are unaffected by the disease process. When considering function, it is important to point out that the specific compliance (lung compliance/functional residual capacity) in the lung in ARF is usually in the normal range, suggesting that the elastic characteristics of the healthy and aerated areas of the lung remain normal (Suter et al. 1978; Gattinoni et al. 1982). Moreover, compartments with a normal ventilation/perfusion (V_A/Q) ratio are commonly found in ARF (Dantzker et al. 1979).

Zone R

Zone R is the zone where positive pressure produces some beneficial effects; the CT scan clearly shows, in both normal subjects during anaesthesia (Brismar et al. 1985) and in ARF patients (Gattinoni et al. 1986), that densities may be cleared following the application of increasing pressure. We found that this process correlated closely with improvement in gas exchange (Gattinoni et al. 1986). Zone R may be an area of atelectasis where the collapsed alveolar units are potentially healthy and only need to be re-expanded. The mechanical equivalent of the recruitment may be considered the knee of the pressure/volume curve of the lung (i.e. the "opening pressure" to which most of the alveolar units previously collapsed have been recruited: Lemaire et al. 1984). Once recruited zone R behaves functionally as zone H.

Zone D

Zone D comprises the true areas of disease characterized by alveolar consolidation and/or vascular obliteration. Lung biopsy material and CT imaging clearly indicate the existence of such areas, in which positive pressure is ineffective. Here gas exchange is completely abolished and the zones represent the anatomical equivalent of the true shunt or the alveolar dead space.

It is evident that gas exchange is accomplished only in zones H and R (once recruited). This suggests that gas exchange in the ARF lung is characterized by true shunt (zone D) and normal V_A/Q (zones H and R), the V_A/Q maldistribution being insignificant (Dantzker et al. 1979). This pattern fits with the model of ARF lung described above. If our interpretation is correct it is evident that in an adult patient with ARF the residual healthy "baby" lung (zone H + R), which may be as little as 20%–30% of a normal adult lung, must support all the respiratory burden imposed by an enhanced adult metabolism.

Gas Exchange in ARF

Gas exchange in the ARF lung takes place only in zones R and H. There are three parameters which may be manipulated to maintain oxygenation: they are the FiO_2,

PEEP and mean airway pressure (MAWP). PEEP and MAWP are the only techniques available to produce recruitment in zone R. It is important to emphasize that there is evidence from the literature that MAWP is the main determinant of oxygenation, independent of the mode of ventilation used (CPAP, IRV, HFJV, HFJV–CPPV, CPPV) (Boros 1979; Berman et al. 1981; Gattinoni et al. 1985). These data are consistent with our ARF model in which the V_A/Q maldistribution is not significant and the lung behaves, functionally and anatomically, as normal and diseased. The actual oxygenation obviously depends on the relative proportion of zone H + R to zone D.

To maintain a normal $PaCO_2$ zone H + R must be "hyperventilated", i.e. the ventilation of alveolar units available for gas exchange (tidal volume/functional residual capacity) may be as high as 10 times normal. The necessity for high ventilatory volumes to control the $PaCO_2$ in a reduced lung volume produces high peak pressures, which are superimposed on the PEEP required to maintain oxygenation. A high FiO_2 with increased peak pressures and specific ventilation characterizes the ventilatory pattern of the healthy areas of the ARF lungs, while the diseased areas are left unaffected. CPPV maximally exploits the residual healthy "baby" lung to maintain a viable gas exchange in an adult patient.

Effects of CPPV on "healthy regions" of the lungs

Unfortunately, an increased FiO_2, high peak pressures and specific hyperventilation are all known factors which may contribute to pulmonary damage. Oxygen toxicity is widely recognized, and both human and animal studies clearly show the potential and real damage of oxygen to healthy lung tissue (Deneke and Fanburg 1982). More controversial is the suggestion that peak pressures greater than 30–40 cm H_2O in healthy lungs are deleterious to the lung parenchyma, producing interstitial and alveolar oedema, increased alveolar surface tension, collapse of peripheral airways and microvascular injury (Webb and Tierney 1974; Wyszogrodsky et al. 1975; Kolobow et al. 1985; Dreyfuss et al. 1985; Woodring 1985). Finally, there is considerable evidence that specific hyperventilation, per se, may cause damage to the healthy lung structures (Mascheroni et al. 1985).

In summary, the factors of potential damage to lung structures, namely FiO_2, peak pressure and hyperventilation, are widely used in the management of ARF.

Rationale for Extracorporeal Support in ARF

The goal of respiratory assistance is to provide adequate gas exchange without contributing to further damage to the lungs. As we have emphasized, adequate oxygenation for the given FiO_2 depends only upon the pressure necessary to recruit zone R. Whatever form of ventilation is used, from apnoea to high-frequency, this amount of positive pressure is mandatory. For adequate carbon dioxide removal, adequate ventilation in severe ARF implies high peak pressures and specific hyperventilation of zones H and R. This is unavoidable, whatever form of respiratory support is applied to the diseased lung. The great advantage of extracorporeal support is that a new lung is added to the patient. This artificial lung shares the respiratory burden with the diseased ARF lung by partially substituting for alveolar capillary function.

ECMO and Extracorporeal Carbon Dioxide Removal (ECCO$_2$R)

ECMO and ECCO$_2$R represent the only two forms of extracorporeal support for adult patients with ARF. Although the common aim of these two techniques is to provide time for the diseased lung to heal, the methods through which this goal is pursued are different.

ECMO, as performed in the NIH multi-centre trial, used high-flow venous arterial bypass to provide as much oxygen as possible. The natural lungs were treated with conventional CPPV. ECCO$_2$R uses low-flow veno-venous bypass primarily to clear carbon dioxide production. The diseased lungs are maintained at rest, with only few tidal volumes (3 or 4 breaths per minute) at limited pressures, to preserve the healthy "baby" lung. The pressure necessary for recruitment of zone R is delivered under static conditions. Most importantly, using this technique completely abolishes high peak pressures and specific hyperventilation.

LFPPV–ECCO$_2$R: Technique

The technique of LFPPV–ECCO$_2$R has been described in detail elsewhere (Pesenti et al. 1981; Gattinoni et al. 1983) and will only be summarized here (Fig. 12.2).

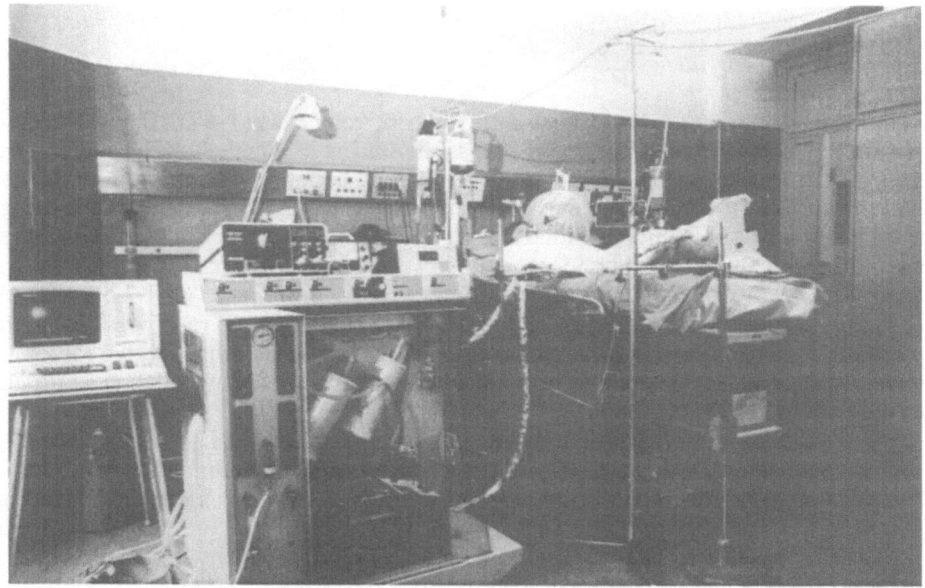

Fig. 12.2.   LFPPV–ECCO$_2$R: clinical set-up.

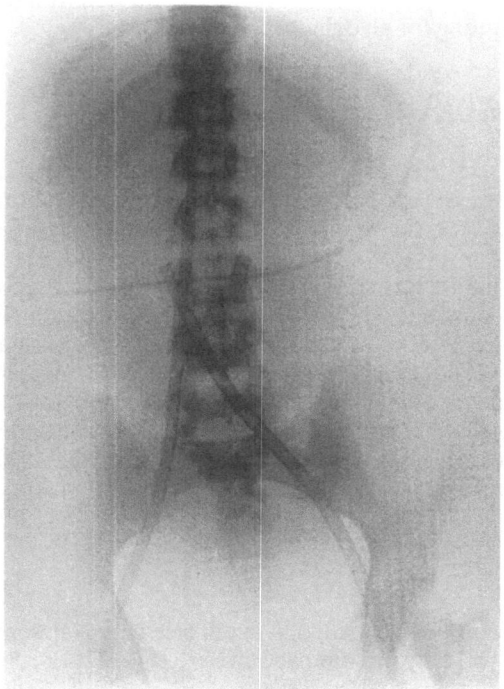

Fig. 12.3. Sapheno–saphenous cannulation.

Extracorporeal Circuit

The drainage catheter is positioned in the inferior vena cava (7–8 cm below the diaphragm). Vascular access is through the saphenous vein (single lumen catheter) or the common femoral vein (double lumen catheter), depending on the vein size. The blood returns to the inferior vena cava 2 cm below the diaphragm through the inner catheter of the double lumen catheter or through a saphenous vein catheter inserted on the other side. The sapheno–saphenous venous cannulation, when feasible, is easier than femoral cannulation, and allows adequate drainage (1.5–2.5 l min^{-1}) (Fig. 12.3). The blood is drained by gravity into a collapsible reservoir and then pumped by a roller pump (1–2 l min^{-1}) through two artificial lungs suitable for long-term use (SCI MED, Kolobow) that are connected in series, before being returned to the patient.

The artificial lungs are ventilated with an air/oxygen mixture (10–15 l min^{-1} each), according to clinical requirements. All the extracorporeal circuit is enclosed in a temperature-controlled compact console (Kontron LSS 6000) and essential monitoring is continuously displayed. This includes the measurement of temperature, blood pressure across the artificial lungs, extracorporeal blood flow and input oxygen saturation. Alarms and feedback controls are provided for gas-blood flows, temperature and pressures. The circuit is provided with suitable ports for an artificial kidney and haemofiltration, if required.

Respiratory circuit

The patient is connected to a ventilator which provides 3 or 4 breaths per minute at limited pressure (35–40 cm H_2O). A small Teflon catheter is advanced through a side port of the tracheal tube to directly above the carina, and provides a supply for oxygen consumption during the end-expiratory pause. The level of positive pressure is set according to the clinical needs.

Clinical Procedure

The patient is paralysed and receives a light intravenous anaesthesia throughout the procedure. Respiratory, haemodynamic and coagulation monitoring is mandatory. After the surgery the patient is connected to the extracorporeal circuit and $ECCO_2R$ begins. As the pulmonary ventilation is decreased to 3 or 4 breaths per minute the PEEP is increased to maintain MAWP at the same level as during the previous CPPV period. The FiO_2 is also maintained at the same concentration. When a consistent improvement in gas exchange has been achieved (usually after a few hours of LFPPV–$ECCO_2R$) the FiO_2 of the ventilator is decreased and the artificial lungs are ventilated with room air. The pressures are decreased when an arterial oxygen partial pressure (PaO_2) higher than 100 mmHg is consistently recorded. Weaning and decurarization are initiated when the shunt fraction is lower than 20%, at an FiO_2 of 0.4, with a total static lung compliance greater than 30 ml (cm $H_2O)^{-1}$ (measured on the inflation limb of the pressure–volume curve, at 10 ml kg $^{-1}$ volume) and when the chest X-ray shows a consistent clearing. The patient is disconnected when viable gas exchange is maintained on CPAP or a low rate of intermittent mandatory ventilation for 6–12 hours without extracorporeal gas exchange.

Clinical Results

Patient Population

The study group consisted of 48 patients (21 males, 27 females) with ARF of various aetiologies (9 bacterial pneumonia, 14 viral pneumonia, 10 post-traumatic, 5 embolism, 10 others). All but two cases were referred to our hospital from other institutions after an average of more than 8 days of mechanical ventilation. To undergo LFPPV–$ECCO_2R$ the patients had to meet the ECMO entry criteria (NHLI 1974) (expected mortality rate higher than 90%) and have total static lung compliance of less than 30 ml (cm $H_2O)^{-1}$. These entry criteria have been fully described and discussed elsewhere (Zapol et al. 1979).

Twenty-five patients (52%) survived and were eventually discharged home. Survivors were obtained from all the aetiological groups. No relationship was observed between survival and the age of the patient. The severity of ARF before the bypass was comparable in survivors and patients who subsequently died, although a significantly lower $PaCO_2$ was observed in survivors at a comparable minute ventilation.

Forty-one patients (85%) had additional organ failure before the bypass. No patient survived failure of five or more organ systems (including the lung). However, survival was over 60% in 12 patients with failure in four systems, including patients in whom both artificial lungs and kidneys were used simultaneously. The mean time on bypass for the survivors was 5.5 days. Irrespective of ultimate survival, 72.9% of the patients had improved lung function during bypass (improved PaO_2, shunt fraction, lung compliance and chest X-ray). The positive response to the bypass therapy was always observed within 48 hours (except in one case). This may have important practical consequences, allowing the withdrawal of bypass if no response is observed within 2–3 days. The patients who did not improve their lung function had significantly higher $PaCO_2$ in the pre-bypass period. This may reflect more advanced lung disease or, alternatively, a more severe injury to the pulmonary microcirculation.

Complications

No major accidents have occurred in more than 8000 hours of extracorporeal perfusion. The average number of organ system failures per patient was not higher during bypass than in the pre-bypass period. In particular, the incidence of sepsis did not increase during bypass. The only adverse finding during the bypass was bleeding. An average of 1800 ± 850 ml per day of whole blood was required to replace the blood losses (including 200–300 ml sampling). Minor bleeding, often requiring surgical revision, occurred at the cannulation sites and major bleeding occurred from chest tubes or during pulmonary surgery while on bypass (3 cases).

Discussion

As outlined above, the main difference between extracorporeal support and any other form of respiratory assistance is that the membrane lung is the only available tool to substitute for the lung's respiratory function. With the use of the membrane lung the healthy regions of the ARF lung may be preserved and not damaged by high pressure and high ventilatory support. It is worth emphasizing again that respiratory assistance with any form of mechanical ventilation is always a symptomatic therapy and does not cure the diseased lung.

Hence our goal is to maintain viable gas exchange without contributing further to lung damage. Two forms of extracorporeal assistance have been used to date in the treatment of the ARF: ECMO and the $ECCO_2R$. Both partially substitute for respiratory function. However, while ECMO was used primarily to provide supplemental oxygen to correct the impairment in oxygenation, $ECCO_2R$ focuses on carbon dioxide removal to abolish the need for ventilation in the diseased lung. ECMO was performed by veno-arterial (v-a) bypass. This may have had some impact on the ultimate survival. During v-a bypass there may be significant pulmonary hypoperfusion. Indeed Ratliff et al. (1975) have suggested that reducing lung perfusion during ARF may lead to pulmonary thrombosis. On the other hand, it is worth remembering that the terminal respiratory units are primarily nourished by pulmonary blood flow. The possibility of lung infarction is always present, and pulmonary infarction was induced experimentally by Kolobow et al. (1981) in healthy sheep after 6 hours of total v-a bypass with the heart in ventricular fibrillation. It may be that v-a bypass decreases the potential for lung repair in the adult patient with ARF.

In contrast, $ECCO_2R$ was always performed using veno-venous bypass, leaving pulmonary and systemic haemodynamics unaffected. However, we believe that the respiratory treatment per se is far more important to the ultimate outcome than the bypass mode. During ECMO the diseased lungs were treated with CPPV. We have discussed in detail the effects of CPPV in ARF lung. During $ECCO_2R$ the LFPPV enables the lungs to rest, thereby avoiding high peak pressures and large minute ventilation. LFPPV was designed to allow a more gentle treatment of the inflamed lung. The reduced mortality rate in patients treated with LFPPV–$ECCO_2R$ compared with CPPV–ECMO may reflect reduced lung damage from LFPPV compared with CPPV.

However, further studies by different centres are required to confirm these data and to substantiate this hypothesis. In our experience we found LFPPV–$ECCO_2R$, although an invasive and cumbersome procedure, to be a safe technique. We have obtained an excellent survival rate in our series of patients with severe ARF who were unresponsive to other conventional or non-conventional forms of respiratory assistance. But it remains to be seen whether, in a randomized controlled trial, LFPPV–$ECCO_2R$ does in fact improve survival.

This work has been supported in part by CNR grant N.840209887 and 85.01487.57, Rome, "Special projects in biomedical engineering".

References

Berman LS, Downs JB, Van Feden A et al. (1981) Inspiration–expiration ratio. Is mean airways pressure the difference? Crit Care Med 9:775–777

Boros SJ (1979) Variations in inspiratory–expiratory ratio and airway pressure wave form during mechanical ventilation: the significance of mean airway pressure. J Pediatr 94:114–117

Brismar B, Hedenstierna G, Lundquist H et al. (1985) Pulmonary densities during anesthesia with muscular relaxation. A proposal of atelectasis. Anesthesiology 62:422–428

Dantzker DR, Brook JC, Dehart P et al. (1979) Ventilation–perfusion distributions in the adult respiratory distress syndrome. Am Rev Respir Dis 120:1039–1042

Deneke SM, Fanburg BL (1982) Oxygen toxicity of the lung: an update. Br J Anaesthesiol 54:737–749

Dreyfuss D, Basset G, Soler P et al. (1985) Intermittent positive pressure hyperventilation with high inflation pressures produces pulmonary microvascular injury in rats. Am Rev Respir Dis 132:880–884

Gattinoni L, Agostoni A, Pesenti A et al. (1980) Treatment of acute respiratory failure with low frequency positive pressure ventilation and extracorporeal CO_2 removal. Lancet II 292–294

Gattinoni L et al. (1982) La compliance toraco-polmonare: metodi di misura, significato clinico, implicazioni terapeutiche. In: Atti Corso Agg. Spec. in Rianimazione e Terapia Intensiva, Pavia, pp 89–104

Gattinoni L, Pesenti A, Kolobow T et al. (1983) A new look at therapy of the adult respiratory distress syndrome: motionless lungs. Int Anaesthesiol Clin 21(2):97–117

Gattinoni L, Pesenti A, Caspani ML et al. (1984) The role of total static lung compliance in the management of severe ARDS unresponsive to conventional treatment. Intensive Care Med 10:121–126

Gattinoni L, Marcolin R, Caspani ML et al. (1985) Constant mean airway pressure with different patterns of positive pressure breathing during the adult respiratory distress syndrome. Clin Respir Physiol 21:275–279

Gattinoni L, Mascheroni D, Torresin A, Marcolin R et al. (1986) Morphological response to positive end expiratory pressure in acute respiratory failure: computerized tomography study. Intensive Care Med 12:137–142

Gille JP, Bagniewski A (1976) Ten years of use of extracorporeal membrane oxygenation (ECMO) in treatment of acute respiratory insufficiency. Trans Am Soc Artif Intern Organs 22:102–108

Hill JD, O'Brien TG, Murray JT (1972) Prolonged extracorporeal oxygenation for acute post traumatic respiratory failure (shock-lung syndrome). N Engl J Med 286:629–634

Kolobow T, Spragg R, Pierce J (1981) Massive pulmonary infarction during total cardiopulmonary bypass in unanesthetized spontaneously breathing lambs. Int J Artif Organs 4:76–81

Kolobow T, Moretti M, Fumagalli R et al. (1985) Adult respiratory distress syndrome following mechanical pulmonary ventilation at high peak airway pressures. Am Rev Respir Dis 131S:137A

Lemaire F et al. (1984) Total respiratory pressure–volume curve in the Adult Respiratory Distress Syndrome. Chest 86:58–66

Mascheroni D, Kolobow T, Fumagalli R et al. (1985) Respiratory failure following induced hyperventilation. An experimental study. Crit Care Med 13:330

NHLI (1974) Protocol for extracorporeal support for respiratory insufficiency. Collaborative Program, National Heart and Lung Institutes, Division of Lung Diseases, 15 May 1974

Pesenti A, Pelizzola A, Mascheroni D et al. (1981) Low frequency positive pressure ventilation with extracorporeal CO_2 removal (LFPPV-ECCO$_2$R) in acute respiratory failure (ARF): technique. Trans Am Soc Artif Intern Organs 27:263–266

Ratliff JL, Hill JD, Fallat RJ et al. (1975) Complications associated with membrane lung support by veno-arterial perfusion. Ann Thorac Surg 19:537–539

Suter MP, Fairley HB, Isemberg MD (1978) Effect of tidal volume and positive end-expiratory pressure on compliance during mechanical ventilation. Chest 73:158–162

Webb H, Tierney DF (1974) Experimental pulmonary oedema due to intermittent positive pressure ventilation with high inflation pressures. Protection by positive end-expiratory pressure. Am Rev Respir Dis 110:556–565

Woodring JH (1985) Pulmonary interstitial emphysema in the adult respiratory distress syndrome. Crit Care Med 13:786–791

Wyszogrodsky I, Kyei-Aboagye K, Taensch HWJR et al. (1975) Surfactant inactivation by hyperventilation: conservation by end-expiratory pressure. J Appl Physiol 38:461–466

Zapol WM, Snider MT, Hill JD et al. (1979) Extracorporeal membrane lung oxygenation in severe acute respiratory failure. JAMA 242:2193–2196

Section IV

Some Aspects of Cardiovascular Support

Chapter 13

The Role of Fluid Replacement in Acute Endotoxin Shock

U. Kreimeier, Zh. Yang and K. Messmer

While it is generally agreed that volume replacement is one of the most important aspects of the treatment of septic shock, the choice of the volume substitute continues to be the subject of controversy. In this chapter we shall review some of the more relevant literature concerning volume replacement and the pathophysiology of septic shock. In addition, new aspects of hyperdynamic endotoxaemia evolving from a recently developed experimental model in the pig are discussed with respect to the effects of volume substitution using either Ringer's lactate or dextran 60.

Pathophysiology of Septic Shock

Macro- and Microcirculatory Failure

In contrast with other forms of shock, septic shock is associated with a wide spectrum of symptoms but characteristically begins with hyperventilation and a high flow state which may progress to a low output/high resistance state (Staub 1974; Villazon et al. 1975). Fig. 13.1 lists the changes in the haemodynamic parameters observed during these two phases of sepsis. These biphasic circulatory changes observed in patients with sepsis were initially described in the 1960s (Clowes et al. 1966; MacLean et al. 1967; Siegel et al. 1967; Udhoji and Weil 1965; Waisbren 1964). Since then, patients have as a result been assigned to different categories of septicaemia and septic shock depending upon their central haemodynamic parameters (high or low central venous pressure and cardiac output, respectively), peripheral vascular tonus (high or low systemic resistance) and metabolic state (acidosis or alkalosis) (Clowes et al. 1966, 1974; MacLean et al. 1967; Shoemaker, 1971; Siegel et al. 1967; Udhoji and Weil

Sepsis	AOP	CVP	VO$_2$	avDO$_2$	TPR	CI
hyperdynamic	↔	↔	(↑)	(↑)	↓	↑
	↓	↔/↑	↓↓	↓↓	↓↓	↑↑
hypodynamic	↓↓	↔/↑	↓	↓	↓↓	↓↓
	↓↓	↔/↑	↑↑	↑↑	↓↓	↓↓

Fig. 13.1. Changes in haemodynamic and ventilatory parameters as the hyperdynamic state of sepsis progresses into a low flow state. AOP, aortic pressure; CVP, central venous pressure; VO$_2$, total body oxygen consumption; avDO$_2$, arteriovenous difference for oxygen; TPR, total peripheral resistance; CI, cardiac index.

1965). However, it seems that circulatory failure in septic shock is most often related to the following characteristic pathophysiological alterations:

1. Reduction of the effective circulating volume due to fluid translocation
2. Relative failure of the right ventricle due to increased pulmonary vascular resistance
3. Impaired function of the left ventricle.

The decrease in cardiac output, *systemic hypotension* and *pulmonary hypertension* probably reflect venous pooling and plugging of capillary beds. The maldistribution of blood flow *within* organs leads to reduced renal blood flow (oliguria), decreased cerebral flow (mental obtundation), a pulmonary capillary leak with a large ventilation–perfusion mismatch and generalized tissue hypoxia (lactic acidosis), which are the prominent features of patients with septic shock who develop multiple organ failure (Baue 1975; Clowes et al. 1975; Pine et al. 1982; Waisbren 1978). Recently, the spectrum of cardiovascular alterations taking place in association with gram-negative sepsis has been further elucidated by Hess (Hess et al. 1981). These authors have stressed in particular the clinical relevance of Waisbren's original description of two different populations of patients: those capable of sustaining a normal or high cardiac index and who are likely to survive, and those who, in contrast, fail to maintain a normal cardiac index and progress into a low flow state and, despite treatment, rarely recover.

Imbalance of Transmembranous Fluid Exchange

In the attempt to improve nutritional perfusion by fluid loading one faces constantly the almost inevitable risk of increasing interstitial pulmonary oedema with a further impairment of pulmonary function (Jardin et al. 1979). It is therefore essential to determine sequentially the actual requirement for fluid replacement.

As stated in the review by Gamble (Chapter 1), extracellular fluid constantly equilibrates in accordance with the intravascular and interstitial hydrostatic and col-

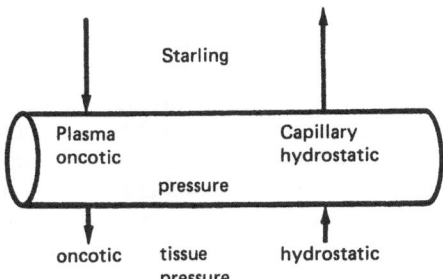

Fig. 13.2. Equilibration of intravascular and interstitial pressures according to the Starling equation. (From Gruber et al. 1977.)

loid osmotic pressures (Fig. 13.2; Gruber and Messmer 1977). Under normal conditions capillary hydrostatic pressure plus tissue oncotic pressure are of equal magnitude to plasma oncotic pressure plus tissue hydrostatic pressure, but of opposite direction. Depending upon the organ and the type of vascular endothelium, the value for the protein reflexion coefficient (σ), which determines the effective transvascular oncotic pressure difference, varies from 0 to 1 and is both molecule- and pore-specific. In other words, each plasma protein molecule, as well as the molecules of polydispersed artificial colloids, have a given reflexion coefficient for each pore found in the endothelial wall (Gabel 1985). For the healthy capillary endothelium of the lungs σ has been found to be 0.8 to 0.9 for albumin (Staub 1974a, b). In animals, the protein concentration in the interstitium is thought to be equivalent to the protein concentration of the lung lymph (Erdmann et al. 1975), which varies from 50 to 80% of the plasma concentration depending upon the species investigated (Staub 1974a, b). The major mechanism for the preservation of a "relatively" dry interstitium is the drainage of fluid via the pulmonary lymphatics from the interstitium to the systemic circulation (Staub 1974a, b).

Staub (1974a, b) has suggested that the *rates* of fluid and protein flow through the pulmonary interstitium and lymph system may increase several-fold before any (clinically measurable) increase in the extravascular fluid *content* takes place (Fig. 13.3). Under normal conditions the interstitial osmotic pressure decreases when the intra-

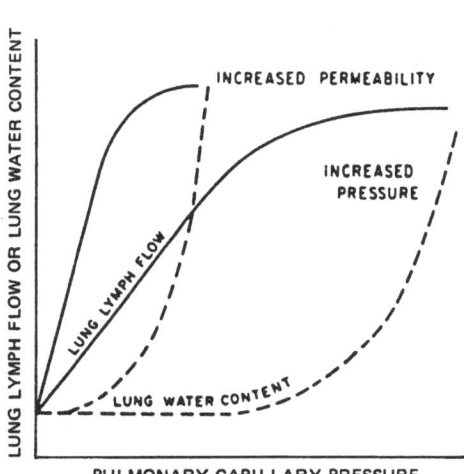

Fig. 13.3. Dependence of lung lymph flow or lung water content upon pulmonary capillary pressure at normal and increased permeability and intravascular pressure, respectively. (From Sibbald 1981.)

vascular hydrostatic pressure increases: thus, in early oedema formation caused by increased pulmonary microvascular hydrostatic pressure, the washout of lung tissue protein with the lung lymph constitutes an important safety factor by increasing the plasma to tissue oncotic pressure difference. However, in the case of augmented permeability as seen in sepsis, the increase in transmembranous movement of water is greater than that during pulmonary oedema secondary to hydrostatic factors alone (Brigham et al. 1979; Sibbald and Driedger 1981), because this mechanism protecting against enhanced extravascular movement of fluid is impaired (Fig. 13.4). In overwhelming sepsis as well as during severe anaphylactic reactions the endothelial wall, especially of the postcapillary venules, becomes leaky with the result that albumin and larger protein molecules such as fibrinogen, but also red blood cells, can pass through the microvascular membranes. Thus, the increased efflux of water is accompanied by an increased efflux of protein (Erdmann et al. 1975; Fleck et al. 1985; Grundmann et al. 1985; Sibbald and Driedger 1981).

The increased transmembranous flow was further investigated by Hill (1980), who injected live *E. coli* bacteria intravenously into dogs which then were overloaded with fluid until the left atrial pressure reached 35 cm H_2O. The increase in extravascular lung water correlated directly with the increase in left atrial pressure, reaching statistical significance at 35 cm H_2O when compared with dogs without sepsis receiving equal amounts of fluid. Furthermore, the relationship between fluid protein concentration and the oncotic pressure is not linear; when the protein concentration decreases in both the plasma and lung interstitium to the same degree, the van't Hoff rule predicts that the net oncotic gradient will decrease, whereby accumulation of water in the pulmonary interstitium is favoured (Hauser et al. 1980). Therefore, in hypoproteinaemia and at low colloid osmotic pressure, volume loss into the interstitium will necessarily increase. When the plasma oncotic pressure decreases,

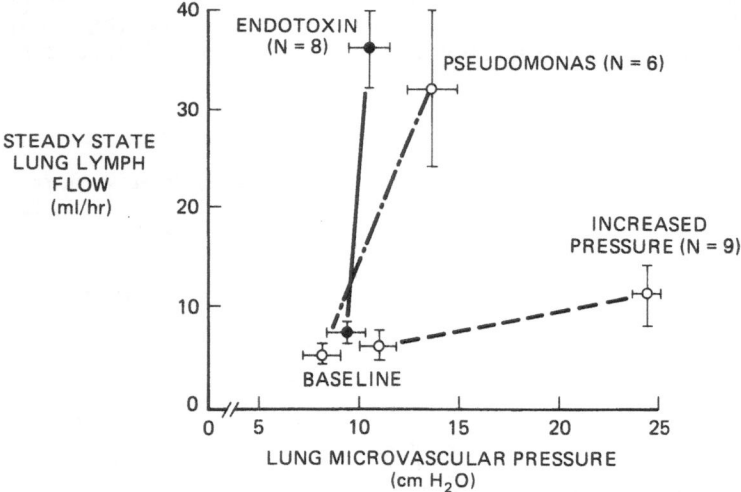

Fig. 13.4. Increase in lung lymph flow during endotoxaemia, bacteraemia (caused by *Pseudomonas*) and as result of increased intravascular pressure. Note that increased pressure per se has little effect on lymph flow. (From Sibbald 1981.)

fluid shift into the interstitium can be further propagated when crystalloids are administered exclusively: even at normal plasma colloid osmotic pressure approximately 80% of crystalloid administered might be expected to enter the interstitium and only 20% of the total amount given will expand the plasma volume (Shoemaker and Hauser 1979).

The Need for Volume Support

According to Siegel (Siegel et al. 1967), patients in septic shock present with an indirect correlation (as plotted in a log diagram) between vascular tone (total peripheral resistance) and cardiac index. Shoemaker (1971) concluded from his investigations that the decrease in cardiac output combined with failure of the circulatory system's functional capacity to deliver oxygen to the tissues is the major determinant of survival. He demonstrated that it was not the degree of arterial hypotension that determines survival but the nutritional flow to the tissues (see Chapter 7). Already in the early phase of peritonitis two causes of inadequate circulation may be present (Clowes et al. 1966; Udhoji and Weil 1965): first, the deficit in intravascular volume accompanying the formation of a large third space, and second, the reduction in venous return and myocardial failure associated with the presence of bacterial toxins in the circulation. Clowes et al. (1966) found an improvement of the circulatory and metabolic state in the majority of their patients as the direct result of intravenous fluid replacement. In the clinical studies of Gunnar et al. (1973), vigorous volume replacement resulted in an increase in cardiac output and a stabilization of the arterial pressure with subsequent survival.

From the many studies of invasive haemodynamic monitoring of patients in septic shock it is now well recognized that the quantity of fluid required to obtain adequate filling pressures may be considerably in excess of the "normal blood volume" and can amount to 8–12 litres within only a few hours. Consequently, the choice of volume substitution method is of the utmost importance. Wiles et al. (1980) judged preload to be maximal when further volume replacement did not produce a significant rise in cardiac output. However, this regimen does not appear suitable and safe for all critically ill patients. The right and left ventricular filling pressures in patients in shock, especially those with left ventricular failure, do frequently dissociate; therefore central venous pressure alone cannot be used as a parameter for fluid management. In contrast, by using a pulmonary artery flotation catheter better information concerning the haemodynamic changes occurring early in the initial phase of septic shock can be derived. Furthermore, independent of the need for balancing intravascular volume and ventricular preload, one has to consider the possibility of myocardial dysfunction, which frequently develops during the course of sepsis (Krausz et al. 1977; Winslow et al. 1973). With the help of on-line monitoring (Swan–Ganz catheter) and adequate fluid administration, Blaisdell (1981) did not observe states of hypodynamic septic shock, except as terminal events.

The rate of oxygen consumption has been suggested as an important measure for assessing the efficacy of therapeutic interventions in sepsis (Houtchens and Westenskow 1984; Shoemaker 1971). Data collected from patients during various stages of septic shock reveal that the probability of survival is higher when oxygen consumption and cardiac output are "supranormal" (Houtchens and Westenskow 1984); this finding is not surprising, as both oxygen availability and calculated oxygen consumption depend upon cardiac output. Following intravenous infusion of live bacteria in monkeys, systemic oxygen consumption declined in parallel with decreasing cardiac output; this fall could be reversed with fluid loading (Carroll and Snyder 1982). Furthermore, rapid administration of fluids was required to produce and maintain the hyperdynamic state in severe bacteraemic sepsis (Carroll and Snyder 1982).

Controversy exists about the importance of changes in plasma colloid osmotic pressure (COP) and the development of pulmonary oedema during sepsis (Kohler et al. 1981; Tonnesen et al. 1977; Staub 1974a, b). In 1985 Sturm et al. (Kant et al. 1985; Sturm et al. 1985) investigated the effects of crystalloid versus colloid therapy (human albumin 4%) on the permeability of the capillary membrane and on the interstitial fluid content of the lung after traumatic haemorrhagic shock. They found that the washout of interstitial protein associated with the high lymph flow as a result of crystalloid therapy was beneficial in terms of interstitial fluid transport. In contrast, colloid therapy had the opposite effect, namely the accumulation of protein molecules in the interstitium hindering fluid drainage. Colloids, when passing *into* the interstitium, augment interstitial oncotic pressure and therefore impair fluid drainage; they were thus considered to have a detrimental rather than a beneficial effect (Sturm et al. 1985).

Grundmann et al. (1985) analysed the effect of an intravenous infusion of 40 g human albumin in 72 patients with or without sepsis. Patients without sepsis showed an increase in the COP as well as in the serum albumin concentration during a 24 hour period. In those with sepsis, however, the serum albumin concentration rose only temporarily and rapidly dropped to the initial values; this disappearance of albumin from the intravascular space indicated sequestration of protein in the extravascular spaces.

Robin et al. (1972) collected pulmonary oedema fluid from two patients with enhanced capillary permeability probably related to circulating endotoxin and compared this fluid with the patients' own plasma. Intravenous administration of dextran 70 as well as of high molecular weight dextran (MW 500 000) led to accumulation of these molecules in the pulmonary oedema fluid at a rate consistent with abnormally high pulmonary capillary permeability. The efficacy of different volume substitutes was stressed in particular by Carroll and Snyder (1982). In their studies on endotoxaemia in monkeys they found it difficult to maintain a hyperdynamic circulatory flow state without the infusion of blood in addition to saline.

Clinical Implications

MacLean et al. (1967) and Clowes et al. (1966) found the presence of the early hyperdynamic state of septic shock closely related to the likelihood of patient survival. Although the mechanisms causing hyperdynamic circulatory shock have not yet been fully elucidated, it is associated with release of various mediators and products from the different endogenous cascade systems (coagulation, complement, arachidonic acid), which per se alter the central as well as the micro-haemodynamics (Messmer 1982, 1985; Oettinger and Seifert 1982). While in the past therapeutic efforts have been directed towards the reversal of changes within the individual organs involved, it seems more promising to establish the diagnosis of hyperdynamic septicaemia and to start general treatment as early as possible.

Hyperdynamic Endotoxaemia: Experimental Data

It was first demonstrated in the experiments of MacLean and Weil (1956) that by intravenous injection of gram-negative endotoxin from *E. coli* a state similar to, if not identical with, the shock state seen in patients with gram-negative sepsis can be produced. Since then it has been confirmed that the constituent of the gram-negative organism responsible for septic shock is endotoxin, a lipopolysaccharide component of the bacterial membrane.

The aim of our experimental studies was to elucidate the role of fluid administration and maintenance of COP in the plasma during the initial phase of acute endotoxaemia. The experimental data were obtained from the pig, which proved to be a suitable animal model for research, in particular because the cardiovascular, respiratory, renal and haematological systems are similar in anatomy and physiological response to those of humans (Dodds 1982).

Methods

Ten barn-raised pigs (body weight 19–27 kg) received a general anaesthesia under controlled ventilation. A basal volume of 100 ml h^{-1} Ringer's lactate was infused throughout the whole experiment. In order to imitate the clinical situation of anaemia occurring during sepsis and in order to gain autologous blood for replacement of blood sampled, the animals were initially haemodiluted by isovolaemic exchange of blood with dextran 60 (Macrodex 6%, Schiwa, D-Glandorf); the haematocrit was lowered to about 20%. After a stabilization interval (30 min) all animals received a continuous infusion of *Salmonella abortus equi* endotoxin (Hermal-Chemie, D-Reinbek) over a period of 210 min.

Fig. 13.5 shows the experimental protocol. The infusion of endotoxin started at point I at a dose of 10 µg $kg^{-1} h^{-1}$. The first change observed was an increase in the pulmonary arterial pressure (PAP); when PAP had stabilized at a plateau value (II), the dose of endotoxin was reduced to 5 µg $kg^{-1} h^{-1}$. At the same time volume substitution was started, and carefully controlled to prevent a fall in pulmonary capillary wedge pressure from its initial control value, by infusion of either dextran 60 (DX) or Ringer's lactate (RL). At 60 and 120 min after point II (III and IV respectively) the dose of endotoxin was further reduced to 2 and 1 µg $kg^{-1} h^{-1}$ respectively. At times I, II, III and IV microspheres were injected into the left ventricle and haemodynamic measurements performed.

The analysis included: central haemodynamics, lung function, gas exchange, the regional organ blood flow (using radioactive labelled microspheres 15 µm in diameter) and the concentration of endotoxin in the plasma (using the chromogenic substrate test; Kabi-Vitrum, S-Stockholm).

In the following results all data are presented as medians and Q_1-/ Q_3-quartiles. Student's paired t-test was used for statistical analysis of differences between baseline and periodically measured data. Kruskal–Wallis one-way analysis of variance was used for comparison of the two treatment groups (DX vs RL); a $p < 0.05$ was considered significant.

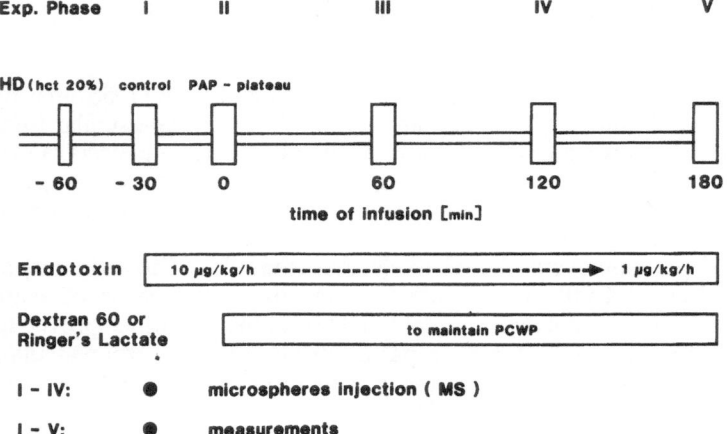

Fig. 13.5. Experimental model of controlled hyperdynamic endotoxaemia using dextran 60 or Ringer's lactate for volume support. PCWP, pulmonary capillary wedge pressure; HD, haemodilution.

Results

The initial response of the arterial, central venous and pulmonary artery pressures to the intravenous infusion of endotoxin in one pig is shown in an original recording in Fig. 13.6.

In both groups (DX, RL) the mean pulmonary arterial pressure (P̄AP) increased to 42 (34/46) mmHg and 46 (46/46) mmHg, respectively. At the plateau of PAP the

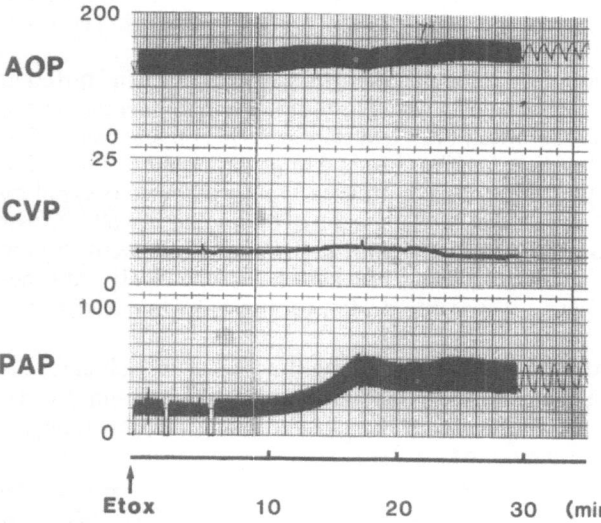

Fig. 13.6. Central haemodynamics in the early phase of acutely induced hyperdynamic endotoxaemia in a pig. AOP, aortic pressure; CVP, central venous pressure; PAP, pulmonary arterial pressure; Etox, endotoxin administration.

calculated pulmonary vascular resistance had also increased to 770 (750/880) and 790 (780/890) dynes s cm^{-5}, respectively. Subsequently the $\bar{P}AP$ declined to 26 (20/29) mmHg and 24 (21/26) mmHg (V), respectively; after 210 min the right ventricular work index was not different from the control value in the DX group, in contrast to a significant fall from 1.1 (1.1/1.2) to 0.7 (0.7/0.8) kg.M/20 kg in RL-treated animals.

The rise in PAP was followed by a decrease in total peripheral resistance from 2300 (2200/2600; control; DX) to 1400 (1100/1700) dynes s cm^{-5} and from 2100 (1900/2600; control; RL) to 1000 (800/1200) dynes s cm^{-5} after 90 min of endotoxin infusion. At the same time there was a precipitous fall in the peripheral white cell count to 1400 (1200/2200) and 1700 (1600/2000) mm^{-3}, respectively. Leucopenia persisted throughout the rest of the observation period.

Despite fluid replacement to maintain left ventricular filling pressure at its initial control value, cardiac index was significantly decreased from 3.8 (3.4/4.9) to 2.1 (2.0/2.6) l min^{-1} (20 kg)$^{-1}$ in the RL-treated animals after 150 min of endotoxaemia, whereas it remained constant in the animals receiving DX as a volume substitute. In contrast to the 28 (23/46) ml kg^{-1} of DX used, 97 (68/110) ml kg^{-1} RL was needed to keep pulmonary capillary wedge pressure constant (Fig. 13.7).

As a consequence of the high crystalloid load COP fell significantly from 28 (27/29) to 12 (11/14) cm H$_2$O in the RL group, whereas it was well maintained in the control range of 27 (23/32) cm H$_2$O in the DX-treated animals.

The concentration of endotoxin in the plasma reached 42 (10/48) ng ml^{-1} in the DX group and 21 (16/23) ng ml^{-1} in the RL group. Ninety minutes after initiating the endotoxin infusion its concentration was 32 (1/73) and 32 (25/65) ng ml^{-1} plasma in both groups. A significant drop in the total protein content of plasma was noted in both groups (DX: 42%; RL: 58% of control values).

Differences between the two groups were not observed in the course of pulmonary ventilatory resistance and compliance nor with regard to oxygen index, which reflected impaired lung function from 90 min of endotoxaemia onwards. At the end of the observation period (210 min endotoxaemia), left cardiac work index and

Parameter	DX	RL
Amount of Endotoxin	identical	
PCWP	constant	constant
CI	constant	↓
TPR	↓	↓
WBC	↓↓	↓↓
COP	constant	↓↓
Fluid volume $\left[\frac{ml}{kg}\right]$	30	100

Fig. 13.7. Characteristic parameters during hyperdynamic endotoxaemia and volume substitution with dextran 60 (DX, $n = 5$) and Ringer's lactate (RL, $n = 5$), respectively. PCWP, pulmonary capillary wedge pressure; CI, cardiac index; TPR, total peripheral resistance; WBC, white blood cell count; COP, colloid osmotic pressure.

Organ	DX	RL
cerebral cortex	constant	constant
right ventricle	↑	↑↑
left ventricle	constant	↓
kidneys	constant	↓
spleen	↓	↓↓
pancreas	constant	↓
gastric mucosa	↓	↓↓
small intestine	constant	↓
colon	constant	↓

Fig. 13.8. Changes in nutritional organ blood flow ($p < 0.05$) during dextran 60 (DX) and Ringer's lactate (RL) volume substitution.

peripheral oxygen availability were significantly reduced to 22% and 55% of their control values, respectively, in the RL-treated animals. This was in contrast to the non-significant changes of these two parameters in the DX-treated animals; however, a greater variation in the data obtained was noted. No difference was seen for arterial oxygen content in either group.

Only sequential changes in regional blood flow (RBF) in vital organs were noted when DX was used for volume substitution. In contrast, in the RL-treated animals RBF reflected impaired nutritional blood flow in most of the organs investigated: blood flows to the left ventricle, kidneys, pancreas, gastric mucosa, small intestine and colon were significantly affected in the RL group. The histological analysis revealed massive oedema of the interstitium at the end of the experiment. The spleen was the only organ in which blood flow was uniformly and progressively diminished independent of the volume substitute. Fig. 13.8 lists the changes observed in RBF of vital organs during hyperdynamic endotoxaemia and volume replacement with the two substitutes investigated.

Summary

A hyperdynamic circulatory state is the characteristic feature of septic shock in its early phase. Survival may be increased if the transition of the high flow state into the hypodynamic, low flow state can be prevented by well-controlled volume support using pulmonary capillary wedge pressure as a readily available clinical guideline.

In our animal study the maintenance of cardiac output in endotoxaemia prevented the development of low flow/high resistance shock. The lungs' ventilatory function was only moderately affected with no significant difference between volume substitution with either dextran 60 or Ringer's lactate. However, we found that vital organs suffer microcirculatory disturbances early in the initial phase of endotoxaemia. Impairment of regional and intra-organ blood flow distribution was more pronounced in the animals treated with Ringer's lactate. Dextran 60 appeared to prevent underperfusion of vital organs by maintaining vascular volume at a constant colloid osmotic pressure; no signs of tissue overhydration were observed.

References

Baue AE (1975) Multiple, progressive or sequential systems failure: a syndrome of the 1970s. Arch Surg 110:779–781

Blaisdell FW (1981) Controversy in shock research: the role of steroids in septic shock. Circ Shock 8:673–682

Brigham KL, Bowers RE, Haynes J (1979) Increased sheep lung vascular permeability caused by *Escherichia coli* endotoxin. Circ Res 45:292–297

Carroll GC, Snyder JV (1982) Hyperdynamic severe intravascular sepsis depends on fluid administration in *Cynomolgus* monkey. Am J Physiol 243:R131–R141

Clowes GHA, Vucinic M, Weidner MG (1966) Circulatory and metabolic alterations associated with survival or death in peritonitis. Ann Surg 163:866–885

Clowes GHA, O'Donnell TF, Ryan NT, Blackburn GL (1974) Energy metabolism in sepsis. Ann Surg 179:684–696

Clowes GHA, Hirsch E, Williams L et al. (1975) Septic lung and shock lung in man. Ann Surg 181:681–692

Dodds WJ (1982) The pig model for biomedical research. Fed Proc 41:247–256

Erdmann AJ, Vaughn TR, Brigham KL (1975) Effect of increased vascular pressure on lung fluid balance in unanesthetized sheep. Circ Res 37:271–278

Fleck A, Raines G, Hawker F, Trotter J, Walace PI, Ledingham McA, Calman KC (1985) Increased vascular permeability: a major cause of hypoalbuminaemia in disease and injury. Lancet I:781–783

Gabel JS (1985) Choices for correct fluid therapy: should there be a controversy? Intensivmed Notfallmed Anaesthesiol 52:18–26

Gruber UF, Messmer K (1977) Colloids for blood volume support. Prog Surg 15:49–76

Grundmann R, Schwarzkopf N, Oette K (1985) Der Einfluss einer postoperativen Humanalbumintherapie auf die Serumeiweissfraktion septischer und aseptischer Patienten. Infusionstherapie 12:246–250

Gunnar RM, Loeb HS, Winslow ES (1973) Hemodynamic measurements in bacteremia and septic shock in man. J Infect Dis 128:5295–5298

Hauser CJ, Shoemaker WC, Turpin I, Goldberg SJ (1980) Oxygen transport responses to colloids and crystalloids in critically ill surgical patients. Surg Gynecol Obstet 150:811–816

Hess ML, Hastillo A, Greenfield LJ (1981) Spectrum of cardiovascular function during gram-negative sepsis. Prog Cardiovasc Dis 23:279–298

Hill SL (1980) Changes in lung water and capillary permeability following sepsis and fluid overload. J Surg Res 28:140–150

Houtchens BA, Westenskow DR (1984) Oxygen consumption in septic shock: collective review. Circ Shock 13:361–384

Jardin F, Eveleigh MC, Gurdjian F, Delille F, Margairaz A (1979) Venous admixture in human septic shock. Circulation 60:155–159

Kant CJ, Sturm JA, Neumann C, Oestern HJ (1985) The capillary membrane and interstitium of the lung during alternative volume therapy after traumatic-hemorrhagic shock. Langenbecks Arch (1985) [Suppl]:75–79

Kohler JP, Rice CL, Yarins CK, Cammack BF, Moss GS (1981) Does reduced colloid oncotic pressure increase pulmonary dysfunction in sepsis? Crit Care Med 9:90–93

Krausz MM, Perel A, Eimerl D, Cotev S (1977) Cardiopulmonary effects of volume loading in patients in septic shock. Ann Surg 185:429–434

MacLean LD, Weil MH (1956) Hypotension (shock) in dogs produced by *Escherichia coli* endotoxin. Circ Res 4:546–556

MacLean LD, Muigan WG, McLean APH, Duff JH (1967) Patterns of septic shock in man: a detailed study of 56 patients. Ann Surg 166:543–558

Messmer K (1982) Pathophysiologie des septischen Patienten. Intensivmed Notfallmed Anaesthesiol 37:12–26

Messmer K (1985) Septic shock: pathophysiology and clinical features. Intensivmed Notfallmed Anaesthesiol 52:2–3

Oettinger W, Seifert J (1982) Pathophysiologische Bedeutung der Prostanoide im septischen Schock. Fortschr Med 100:2169–2174

Pine RW, Wertz MJ, Lennard ES, Dellinger EP, Carrico CJ, Minshew BH (1982) Determinants of organ malfunction or death in patients with intra-abdominal sepsis. Arch Surg 118:242–249

Robin ED, Carey LC, Grenvik A, Glauser F, Gaudio R (1972) Capillary leak syndrome with pulmonary edema. Arch Intern Med 130:66–72

Shoemaker WC (1971) Cardiorespiratory patterns in complicated and uncomplicated septic shock. Ann Surg 174:119–125

Shoemaker WC, Hauser CJ (1979) Critique of crystalloid versus colloid therapy in shock and shock lung. Crit Care Med 7:117–124

Sibbald WJ, Driedger AA (1981) Pulmonary alveolar capillary permeability in human septic respiratory distress syndrome. In: Cowley RA, Trump BF (eds) Pathophysiology of shock, anoxia and ischemia. Williams and Wilkins, Baltimore, pp 372–387

Siegel JH, Greenspan M, Del Guercio LRM (1967) Abnormal vascular tone, defective oxygen transport and myocardial failure in human septic shock. Ann Surg 165:504–517

Staub NC (1974a) "State of the art" review. Pathogenesis of pulmonary edema. Am Rev Respir Dis 109:358–372

Staub NC (1974b) Pulmonary edema. Physiol Rev 54:678–811

Sturm JA, Creutzig H, Oestern HJ, Maghsudi M, Wisner DH, Schober O (1985) Albumin extravasation as a method of following pulmonary permeability changes in multiple-trauma patients. Langenbecks Arch (1985) [Suppl]:69–73

Tonnesen AS, Gabel JC, McLeavey CA (1977) Relation between lowered colloid osmotic pressure, respiratory failure, and death. Crit Care Med 5:239–241

Udhoji VN, Weil MH (1965) Hemodynamic and metabolic studies on shock associated with bacteremia. Ann Int Med 62:966–978

Villazon SA, Sierra UA, Lopez SF, Rolando A (1975) Hemodynamic patterns in shock and critically ill patients. Crit Care Med 3:215–221

Virgilio RW, Rice CL, Smith DE, James DR, Yarins CK, Hobelmann CF, Peters RM (1979) Crystalloid vs. colloid resuscitation: is one better? Surgery 85:129–139

Waisbren BA (1951) Bacteremia due to gram negative bacilli other than *Salmonella*. Arch Intern Med 88:467–488

Waisbren BA (1964) Gram-negative shock and endotoxin shock. Am J Med 36:819–824

Waisbren BA (1978) A paradigm that explains gram-negative shock. Am J Med 65:403–405

Wiles JB, Cerra FB, Siegel JH, Border JR (1980) The systemic septic response: does the organism matter? Crit Care Med 8:55–60

Winslow EJ, Loeb HS, Rahimtoola SH, Kamath S, Gunnar R (1973) Hemodynamic studies and results of therapy in 50 patients with bacteremic shock. Am J Med 54:421–433

Chapter 14

Optimal Use of Vasoactive Agents in Septic Shock

J. L. Vincent

For many years the treatment of septic shock essentially concentrated on the restoration of systemic blood pressure. Metaraminol, mephentermine, noradrenaline, angiotensin and other vasoconstrictors were widely used with rather limited success. More recently, dopamine has become the agent of choice as a vasopressor in the management of circulatory shock. There are several reasons for this. In the first place it combines alpha- and beta-mimetic properties of various strengths according to the dose administered, so that the drug can easily be titrated according to the clinical status of the patient. Secondly, it induces tachyarrhythmias less frequently than other available catecholamines. Finally, it can selectively dilate renal and mesenteric vascular beds when low doses are used.

After more than 20 years of use, there is no substantial evidence that dopamine has improved outcome from circulatory shock, and for this reason other options are still required. More recently, the concept that restoration of arterial pressure is less important than the optimization of blood flow and cellular metabolism has become the underlying principle in the management of shock. Although vasoconstrictors may be indicated in some hyperkinetic states, they should be avoided whenever possible in other clinical conditions. On the other hand, the use of agents with vasodilating properties, such as dobutamine, has been recently suggested for improving tissue perfusion in shock states.

Optimization of blood flow also necessitates initial infusion of fluid, since fluid deficits are almost always present in acute circulatory failure. Fluid therapy has probably improved with the development of sophisticated cardiovascular monitoring and the application of stricter criteria. Nevertheless, fluid administration can be hindered by alterations in vascular compliance and especially changes in myocardial contractility. Obviously, an excessive increase in cardiac filling pressures must be avoided to limit the formation of pulmonary oedema, especially in the presence of altered capillary permeability related to sepsis.

Therefore, intravenous fluids and vasoactive drugs are commonly combined in the management of septic shock and these two forms of therapy should be considered together.

High Flow States

Septic shock is characterized by a distributive defect typically manifested by a high cardiac output and low systemic vascular resistance. This so-called hyperdynamic state is associated with decreased tissue oxygen extraction. There is a continuing controversy regarding the existence of either an anatomical shunt or a metabolic shunt accounting for the reduced oxygen extraction. According to the first theory, blood would be diverted away from cells by the existence of arteriovenous communications. According to the second theory, a cellular defect in oxygen extraction might occur and result in a reduction in oxygen extraction. The two mechanisms could both be involved in the disturbance of oxygen transport since both distributive defects and early cellular metabolic defects have been demonstrated in septic shock (Chaudry 1983).

Generalized vasodilation associated with the hyperdynamic state of septic shock is in part related to the liberation of various substances including endorphins, histamine, bradykinin and prostacyclin. Although vasoconstrictors such as catecholamines, serotonin, angiotensin and thromboxane are also released, the net effect may be a decreased vascular tone. Also important is a decreased responsiveness of the vascular system to alpha-1 adrenergic stimulation, which is related to several factors, including down-regulation of alpha-1 adrenergic receptors, an endotoxin-induced decrease in adrenergic vascular effects and the liberation of interfering substances such as eicosanoids or opioids. These factors have recently been reviewed in more detail elsewhere (Chernow and Roth 1986; Parrillo 1986).

Various pharmacological approaches have been introduced in the management of these particular conditions, including substances interfering with eicosanoids and opioids. Among the catecholamines, dopamine has been most widely used at relatively high doses to counter the vasodilation. Noradrenaline has also been used, sometimes in combination with low doses of dopamine to maintain renal blood flow. The use of calcium agonists such as BAY 8644 may be very valuable, since by increasing cellular calcium availability these substances could enhance both myocardial contractility and vascular reactivity (Schramm et al. 1982). In the past it was a widely held view that hyperdynamic sepsis had a better prognosis than the hypodynamic state, but this is probably not the case. Hyperdynamic septic states are not associated with an improved overall survival (Parker et al. 1984; Azimi and Vincent 1987). However, the haemodynamic state is usually more stable in these conditions. Other than in agonal states or in the case of underlying cirrhosis (Baumgartner et al. 1984), circulatory collapse is usually not so profound and the requirement for vasopressors is therefore relatively limited.

Myocardial Depression

The early haemodynamic studies of human septic shock separated this condition into two distinct entities: the first, a high-flow, low-resistance form of "warm" shock, and the second, a low-flow, high-resistance form of "cold" shock. It was usually thought that myocardial competence was unaltered in the first condition. However, Parker and co-workers have shown by radionuclide techniques that myocardial depression can be characterized by significant ventricular dilatation associated with a decreased left ventricular ejection fraction (Parker et al. 1984). Other studies have confirmed that abnormal left ventricular ejection fraction cannot be predicted from standard haemodynamic measurements (Ellrodt et al. 1985). Furthermore, cardiac dysfunction in septic shock is often a segmental rather than a global process (Ellrodt et al. 1985). It is true to say that myocardial depression is an almost universal feature of septic shock. Several causes have been suggested, including decreased coronary perfusion, myocardial oedema and liberation of cardiodepressant substances. Recent studies indicated that coronary perfusion is usually unaltered in septic shock (McDonough et al. 1985). The release of a myocardial depressant factor, initially suggested by Lefer in 1970 (Lefer 1970), has been a matter of considerable debate. However, Parrillo and co-workers have also now demonstrated the presence of a cardiodepressant substance in the blood of subjects with sepsis (Parrillo et al. 1985). Their assay system was based on the electronic evaluation of the motion of beating rat myocardial cells in the presence of human serum from patients with sepsis and from controls.

Low Flow States

If a high flow does not exclude the presence of underlying myocardial depression, a low cardiac output is not synonymous with depressed contractility since preload, afterload and heart rate are other determinants of cardiac output. In septic shock cardiac preload may be reduced as a consequence of fluid losses and vascular pooling of blood associated with the distributive defect. Increased ventricular afterload can also contribute to decreased blood flow. In particular, pulmonary hypertension, which is frequently observed in sepsis, can increase right ventricular afterload and contribute to right ventricular failure (Clowes et al. 1970; Vincent et al. 1981). Decreased heart rate is relatively uncommon, but can occur in patients with cardiac electrophysiological abnormalities or in those previously treated with beta-blocking agents.

In our experience the hypokinetic state in septic shock is most usually related to reduced cardiac preload and/or myocardial depression. Nevertheless, in patients with septic shock due to peritonitis, fluid administration alone produces a marked increase in stroke volume along with plasma volume in patients who survive, whereas in those who subsequently die, volume loading leads to signs of severe heart failure (Vincent et al. 1981). The use of adrenergic agents in these very septic conditions is of limited value (Vincent et al. 1981).

Choice of Vasoactive Agents

Obviously, the use of vasoconstrictors may be indicated when arterial hypotension is so severe that coronary perfusion is markedly compromised. As soon as this critical

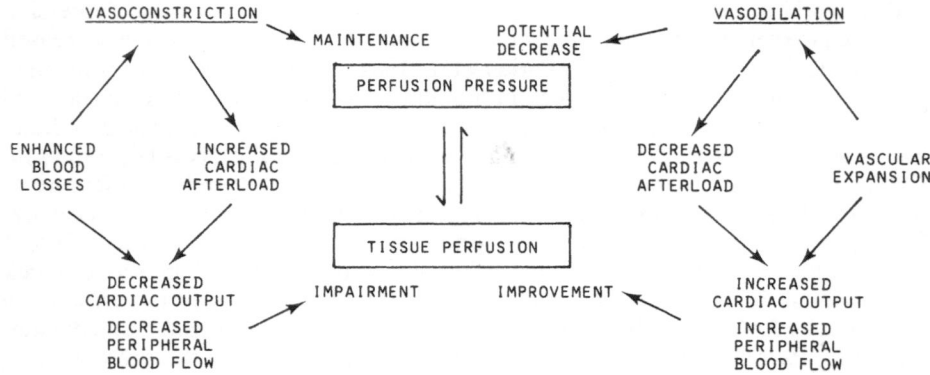

Fig. 14.1. The dilemma created by cardiovascular alterations in sepsis. Vasoconstriction can maintain perfusion pressure but could ultimately impair tissue perfusion, while vasodilation can decrease perfusion pressure but is liable to increase peripheral blood flow.

perfusion pressure is restored, however, there is no clear indication that maintenance of blood pressure can decrease morbidity or mortality from septic shock. There is more evidence that vasoconstrictive agents do harm by decreasing microcirculatory blood supply and increasing cardiac afterload (Fig. 14.1). In some conditions a positive inotropic action alone can also increase arterial pressure (Fig. 14.2).

Although dopamine has combined alpha- and beta-adrenergic and dopaminergic activity, it can induce vasoconstriction even at relatively low doses, for two reasons.

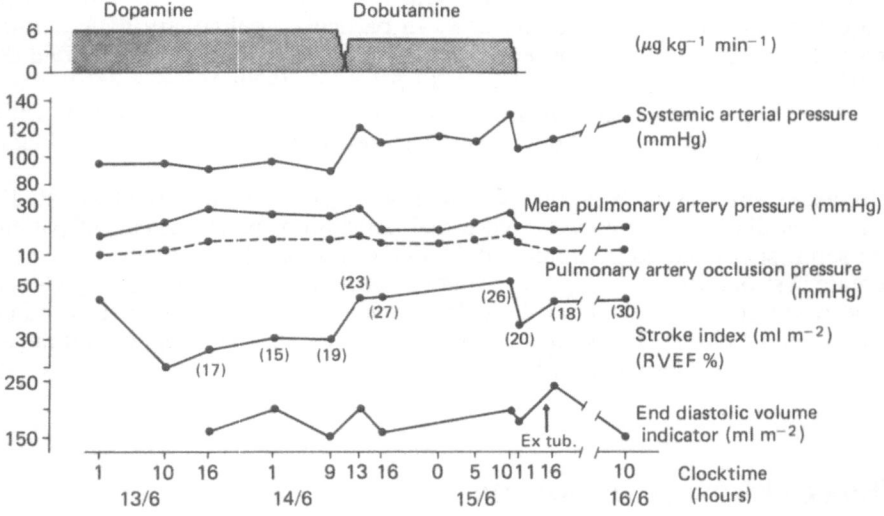

Fig. 14.2. Haemodynamic effects of successive administration of dopamine and dobutamine during septic shock of urinary source (after endoscopic prostate resection) in an 82-year-old patient. Heart rate was around 90 per minute throughout the specific course. Dobutamine markedly improved left and right ventricular function in this particular patient.

Firstly, beta-adrenergic receptors are rapidly down-regulated (Bristow et al. 1982) and alpha-adrenergic effects become predominant. Secondly, dopamine acts in part by the liberation of endogenous noradrenaline which has predominantly a vaso-constrictive action. Dopamine can also increase cardiac filling pressures and thereby limit fluid administration and favour pulmonary oedema formation. It produces a greater increase in cardiac filling pressures than does dobutamine in heart failure (Leier et al. 1978), in cardiogenic shock (Francis et al. 1982), after open heart surgery (Tyden and Nystrom 1983), and in acute respiratory failure (Molloy et al. 1984). When comparing dopamine and dobutamine administration in human septic shock, Regnier and co-workers also observed a fall in filling pressures with dobutamine but an increase with dopamine (Regnier et al. 1979).

Taking a slightly different approach, we investigated the combined effects of fluid administration with either dopamine or dobutamine during septic shock induced by endotoxin infusion in dogs (Domb et al. 1985). To reproduce the usual clinical management of septic shock, we titrated the fluid infusion to maintain pulmonary artery balloon-occluded pressure at a constant level, without drug or with dopamine or with dobutamine at 6 µg kg^{-1} min^{-1}. The amount of fluid infused was significantly greater with dobutamine than with dopamine. This combination of dobutamine with fluid resulted in a significantly higher cardiac output. Oxygen consumption was also higher, indicating that tissue oxygen availability was increased. A high oxygen consumption is undoubtedly beneficial in septic shock (Abraham et al. 1984).

Comparable haemodynamic effects can be obtained with other agents with combined inotropic and vasodilating properties such as the recently available dopexamine or non-adrenergic compounds such as amrinone (Domb et al. 1986). Indeed, we have recently observed a remarkable increase in oxygen transport and oxygen consumption with amrinone during experimental septic shock (Domb et al. 1986). By decreasing capillary hydrostatic pressure, vasodilators can also facilitate an increase in plasma volume by increasing endogenous fluid shifts from the interstitium into the vascular space (Gustafsson and Lundvall 1984). Since fluid therapy is a cornerstone in the treatment of septic shock, and since myocardial depression can commonly hinder fluid replacement by increasing cardiac filling pressures, the effects of the combination of fluids and vasoactive agents in the management of circulatory shock need to be examined. After all, that is what is done in everyday practice.

References

Abraham E, Bland RD, Cobo JC, Shoemaker WC (1984) Sequential cardiorespiratory patterns associated with outcome in septic shock. Chest 85:75–80

Azimi G, Vincent JL (1986) Long term survival from septic shock: Is there a typical hemodynamic pattern? Resuscitation 14:245–253

Baumgartner JD, Vaney C, Perret C (1984) An extreme form of hyperdynamic syndrome in septic shock. Intensive Care Med 10:245–249

Bristow MR, Ginsburg R, Minobe W et al. (1982) Decreased catecholamine sensitivity and beta-adrenergic-receptor density in failing human hearts. N Engl J Med 307:205–211

Chaudry IH (1983) Cellular mechanisms in shock and ischemia and their correction. Am J Physiol 245:R117–R134

Chernow B, Roth BL (1986) Pharmacologic manipulation of the peripheral vasculature in shock: clinical and experimental approaches. Circ Shock 18:141–155

Clowes GH, Farrington GH, Zuschneid W, Cossette GR, Saravis C (1970) Circulating factors in the etiology of pulmonary insufficiency and right heart failure accompanying severe sepsis (peritonitis). Ann Surg 171:663–678

Domb M, Vincent JL, Azimi G et al. (1985) Dopamine versus dobutamine in septic shock: relevance to intravenous fluid administration. Crit Care Med 134:316

Domb M, Van Der Linden P, Azimi G et al. (1986) Treatment of septic shock with amrinone: an experimental study. Crit Care Med 14(4):347

Ellrodt AG, Riedinger MS, Kimchi A et al. (1985) Left ventricular performance in septic shock: reversible segmental and global abnormalities. Am Heart J 110:402–409

Francis GS, Sharma B, Hodges M (1982) Comparative hemodynamic effects of dopamine and dobutamine in patients with acute cardiogenic circulatory collapse. Am Heart J 103:995–1000

Gustafsson D, Lundvall J (1984) B2-adrenergic vascular control in hemorrhage and its influence on cardiac performance. Am J Physiol 246:H351–H359

Lefer AM (1970) Role of a myocardial depressant factor in the pathogenesis of circulatory shock. Fed Proc 29:1836–1847

Leier CV, Heben PT, Huss P, Bush CA, Lewis RD (1978) Comparative systemic and regional hemodynamic effects of dopamine and dobutamine in patients with cariomyopathic heart failure. Circulation 58:466–475

McDonough KH, Lang CH, Spitzer JJ (1985) The effect of hyperdynamic sepsis on myocardial performance. Circ Shock 15:247–259

Molloy WD, Dobson K, Girling L, Greenberg ID, Prewitt RM (1984) Effects of dopamine on cardiopulmonary function and left ventricular volumes in patients with acute respiratory failure. Am Rev Respir Dis 130:396–399

Parker MM, Shelhamer JH, Bacharach SL et al. (1984) Profound but reversible myocardial depression in patients with septic shock. Ann Intern Med 100:483–490

Parrillo JE (1986) Cardiovascular dysfunction in humans with septic shock. In: Vincent JL (ed) Update in intensive care and emergency medicine, Vol 1. Springer-Verlag, Berlin Heidelberg New York, pp 265–274

Parrillo JE, Burch C, Shelhamer JH, Parker MM, Natanson C, Schuette W (1985) A circulating myocardial depressant substance in humans with septic shock. J Clin Invest 76:1539–1553

Regnier B, Safran D, Carlet J, Teisseire B (1979) Comparative haemodynamic effects of dopamine and dobutamine in septic shock. Intensive Care Med 5:115–120

Schramm M, Thomas G, Towart R, Frankowiak G (1982) Activation of calcium channels by novel dihydropyridines: a new mechanism for positive inotropic agents. Nature (Lond) 303:535–537

Tyden H, Nystrom SO (1983) Dopamine versus dobutamine after open-heart surgery. Acta Anesthesiol Scand 27:193–198

Vincent JL, Weil MH, Puri V, Carlson RW (1981) Circulatory shock associated with purulent peritonitis. Am J Surg 142:262–270

Chapter 15

Mechanical Assistance in the Treatment of Shock

C. A. Marshall and W. Kox

The use of mechanical assistance devices in the management of shock is a subject which has received serious attention both in the literature and in clinical practice for two decades. The attraction of these techniques lies in their ability to improve haemodynamic status while minimizing the use of inotropes with their often detrimental effect on myocardial oxygen consumption. The current status of intra-aortic balloon counterpulsation and pneumatic anti-shock trousers is reviewed below.

Intra-Aortic Balloon Counterpulsation

The concept of counterpulsation was first put into practice by Clauss et al. in 1961; a volume of blood was aspirated and re-injected into the aorta of dogs, triggered by the electrocardiogram (ECG). This model had the major problems of haemolysis and considerable inertia. In 1962 Moulopoulos et al. inserted a latex balloon into the descending aorta of dogs which was inflated with carbon dioxide and triggered by the R-wave of the ECG to inflate during diastole and deflate during systole. This was put into clinical practice in 1968 by Kantrowitz et al., since when intra-aortic balloon counterpulsation (IABCP) has been practised worldwide. It has been estimated that by 1981 over 100 000 patients had been treated by the method (Tobias 1981).

IABCP aims to reduce left ventricular work and therefore myocardial oxygen consumption, and also to increase myocardial blood flow. Deflation of the balloon at the instant of isovolumetric left ventricular contraction reduces afterload and improves ejection fraction of the left ventricle. Myocardial oxygen consumption, which depends mainly on the total integrated pressure of the left ventricle and its duration developed during systole (time-tension index) (Saarnof et al. 1958; Braunwald 1971), may therefore be reduced. The improved ejection fraction will then reduce left ventricular end-diastolic volume (preload). In the presence of a competent aortic valve diastolic inflation raises aortic diastolic root pressure. In normal coronary arteries this would not necessarily alter coronary blood flow due to autoregulation. However,

Fuchs et al. (1983) demonstrated an increased blood flow to the left ventricle during IABCP in 7 patients with greater than 90% stenosis of the left anterior descending coronary artery. These increases correlated with an increase in mean aortic diastolic pressure indicating probable loss of autoregulatory ability. Patients with coronary artery disease should therefore benefit from an increased aortic diastolic pressure.

The technique depends on accurate timing of the inflation and deflation of the balloon; early inflation or late deflation will produce an increased left ventricular afterload and therefore increased work. Accurate timing can be achieved by observing the arterial waveform. Balloon inflation is adjusted to occur 40–50 ms before the dicrotic notch (i.e. aortic valve closure) to allow for a phase delay in propagating the arterial waveform. Balloon deflation should occur just before the aortic valve opens (Fig. 15.1). Unloading of the left ventricle will produce an arterial trace where the end-diastolic pressure with pumping is less than that without and the systolic pressure immediately after an assisted beat is lower than that after a non-assisted beat. Ideally, diastolic augmentation (inflation) will produce a diastolic pressure which is greater than the systolic pressure on the arterial trace. Timing has to be adjusted if the heart rate changes significantly.

Applied Physiology

The beneficial effects of IABCP have been attributed to decreased myocardial oxygen demand due to a reduction in left ventricular work, and an increased myocardial oxygen supply due to an increased mean aortic diastolic pressure. The relative importance of each factor is likely to vary with the pathophysiological circumstances. Reduction in left ventricular work should be a major factor in left ventricular failure with normal or increased afterload; diastolic augmentation should be a major factor in the presence of ischaemia.

Studies in dogs and man seem to reflect these variable influences. Spotnitz et al. (1969) demonstrated a reduction in left ventricular wall stress, an increase in coronary blood flow and a decrease in myocardial oxygen consumption during IABCP in normal dogs and those with pharmacological myocardial depression. Powell et al. (1970), however, demonstrated an increase in myocardial oxygen consumption (MVO_2) associated with the increased coronary blood flow (CBF) in hypotensive dogs. Leinbach et al. (1971) recorded CBF and MVO_2 on 14 occasions in patients with cardiogenic shock following myocardial infarction. With IABCP, CBF fell in 7 instances, was unchanged in 3 and rose in 4. Changes in MVO_2 were closely correlated with changes in CBF. Lactate extraction and arterial oxygen tension were not significantly affected. The authors concluded that the net result of IABCP on coronary flow and metabolism depends on the interplay between increased blood flow to ischaemic areas provided by diastolic augmentation and reduction in blood flow to normal myocardium in which the oxygen requirements are reduced by diminished afterload.

Cardiogenic Shock

The use of IABCP in a patient in cardiogenic shock will often produce an immediate improvement in haemodynamics (Ehrich et al. 1977), but this is not always followed

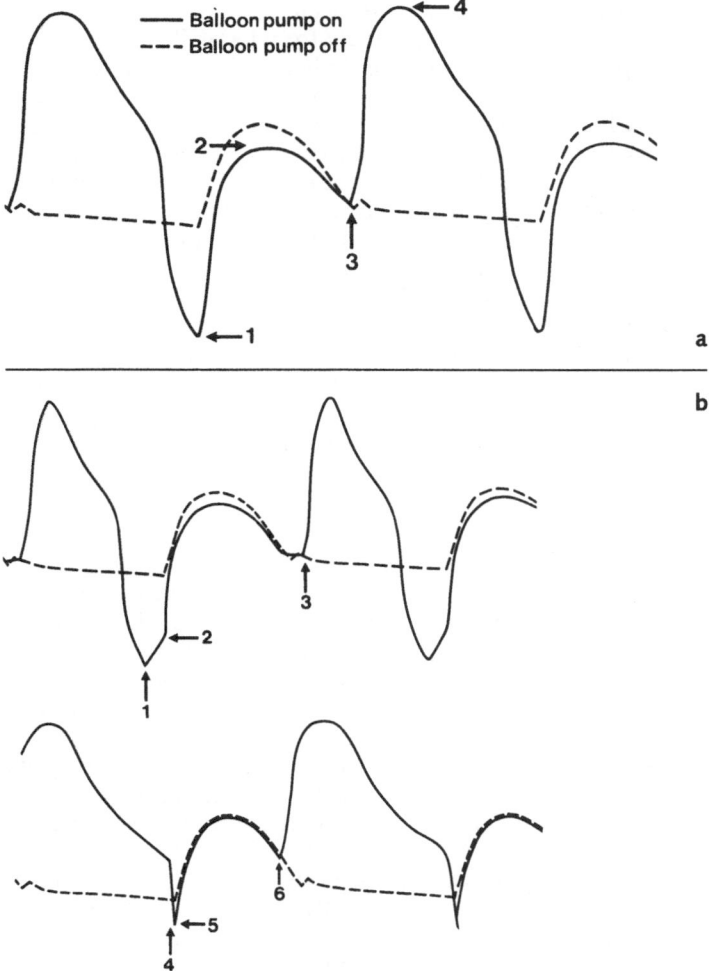

Fig. 15.1. Balloon inflation and deflation. **a** Correct procedure. 1, Aortic valve opening, reduced after-load to left ventricle; 2, peak systolic pressure lower with IABCP; 3, balloon inflation timed to coincide with aortic valve closure; 4, suprasystolic diastolic augmentation. **b** Incorrect procedure. 1, Balloon de-flated too early; 2, aortic valve opening; 3, late inflation of balloon after dicrotic notch; 4, balloon deflated too late leading to increased left ventricular afterload; 5, aortic valve opening; 6, early inflation of balloon.

by an improvement in patient survival. Low survival rates have been reported in several studies: 11% of 18 patients reported by Willerson et al. (1975), 24% of 80 patients in a series by Leinbach et al. (1973), and 17% of 87 patients studied by Scheidt et al. (1973). In this last series only 8 of the original 87 patients survived for more than one year. However, Hagemeijer et al. (1977) found that 14 of 25 patients treated with IABCP survived 3 or more months after an acute infarction complicated by serious pump failure. Attempts have therefore been made to identify the subgroup of patients in cardiogenic shock who will benefit from IABCP.

It is not surprising that patients whose pre-IABCP haemodynamic status is very poor fare less well. Weber et al. (1973) identified a subset of patients in cardiogenic shock who had a pre-IABCP cardiac index of less than $2 \, l \, min^{-1} \, m^{-2}$ and a pulmonary capillary wedge pressure of greater than 15 mmHg. All of these patients died and at post-mortem they were shown to have lost more than 40% of their ventricular muscle mass. A higher survival rate for patients in cardiogenic shock after an inferior myocardial infarction has been noted (Willerson et al. 1975). Alcan et al. (1984) found that balloon counterpulsation coupled with medical therapy alone provided inferior results to surgical management, and this experience has been reported by other groups (Resnekov 1977; McEnany et al. 1978). Alcan et al. (1984) also noted a difference in survival depending on speed of institution of IABCP; 31 of 45 patients in whom IABCP was instituted within 24 hours of the onset of shock left hospital, compared with only 4 of 39 patients in whom IABCP was started after more than 24 hours of shock. Cardiogenic shock remains a condition with a high mortality.

Septicaemic Shock

In preparations of isolated perfused hearts there is progressive deterioration in left ventricular function following an injection of E. coli, with a rise in left ventricular end-diastolic pressure and depressed contractility (Hinshaw et al. 1970; Hinshaw 1974). Patients dying with endotoxic shock have myocardial performance curves that have a lower gradient, lower maximal left ventricular stroke work and lower cardiac index (Weisel et al. 1977).

Dunn et al. (1974) instituted assisted circulation in dogs using a roller pump capable of delivering pulsatile flow; the dogs were given injections of endotoxin (E. coli). Survival was prolonged in animals receiving assisted circulation compared with those which were not. Cardiac output and blood pressure were higher in assisted animals; they also tended to have less metabolic acidosis. Roberts et al. (1979) investigated the cardiovascular effects of septicaemia after administration of Klebsiella aerogenes to 20 dogs, 10 of which received assisted circulation. After 24 hours the left ventricular stroke work was better preserved in assisted animals (minus 25.4% of baseline value) compared with the non-assisted animals (minus 50.1% of baseline value). The assisted animals developed less metabolic acidosis than the non-assisted animals, the arterial pH and lactate staying at baseline. They also noted that after the initial fall in circulating leucocytes seen in all animals there was a later increase in the assisted animals only. Less microscopic injury was observed in the liver and small bowel of assisted animals. The authors suggested that IABCP in sepsis might allow more time for specific therapy to be instituted before irreversible cellular destruction supervenes.

Berger et al. (1973) reported the successful use of IABCP in 2 patients with septicaemic shock and ischaemic heart disease who were unable to develop an increased cardiac output. Mercer et al. (1981) studied the use of IABCP in a patient with hypodynamic septic shock. The technique led to an improvement in cardiac index and the patient ultimately survived. The authors suggest that IABCP may be useful in patients with hypodynamic septic shock who deteriorate despite the administration of antibiotics and vasopressors. Its value in hyperdynamic shock remains doubtful. The literature is limited in this field and there is room for more studies.

Complications

Institution of IABCP is not without hazard; the reported complication rate associated with its use varies between 9% and 24.4% (Goldman et al. 1982; Alcan et al. 1984; Sanfelippo et al. 1986). In addition, in large studies there may be a mortality rate of up to 2% directly attributable to IABCP insertion. This reflects the use of an invasive procedure in patients with vascular disease who may be shocked and septic. The major complications were most commonly vascular problems (limb, renal and mesenteric ischaemia) and sepsis. A post-mortem study of 45 patients (Isner et al. 1980) who died after insertion of an IABCP found a complication rate of 36%; these included aortic dissection, arterial perforation, arterial thrombi or emboli, limb ischaemia and local wound infection. Of the entire 20 complications, only 4 (20%) were suspected before death, indicating that clinical evaluation of complications may underestimate their frequency. In particular, of the 9 cases of arterial dissection none were diagnosed or suspected before post-mortem. This is particularly interesting in the light of a recent paper from Sanfelippo et al. (1986) who performed late follow-up of 283 patients and found that 29% had claudication symptoms in the limb that had hosted the balloon pump. Recent reports (Pennington et al. 1983; Goldberger et al. 1986) have noted a higher complication rate associated with the percutaneous method of IABCP insertion compared with the surgical method.

Pneumatic Anti-Shock Trousers

Pneumatic anti-shock trousers (PTs) have gained wide acceptance in the United States; they are used extensively by ambulance attendants and in hospitals (McSwain 1977; Hoffman 1980). In 1977 the American College of Surgeons listed them as essential equipment in the ambulance (Committee on Trauma 1977). Their use is becoming more widespread in Canada (Butson 1983). Interest was generated during the Vietnam war when Cutler and Daggett (1971) used them to control haemorrhage caused by trauma to the lower limb, pelvis and perineum usually inflicted by land mines; 4 of 8 patients survived after early application of the garment in the field.

Description

The trousers contain inflatable compartments for providing pressure to the limbs and abdomen (Baxter et al. 1984). The garment is applied by lying the patient on it and securing each compartment around the lower limb and abdomen below the costal margin. The compartments are fastened with Velcro. Each compartment can be inflated or deflated individually and is accompanied by either a safety valve that releases pressure at approximately 100 mmHg or a gauge that monitors compartmental pressure. There is an opening at the perineum to allow rectal examination and bladder catheterization. The newer models allow better access to femoral blood vessels and visualization of skin through clear plastic. The trousers are radiolucent and can be left in position while radiography is undertaken.

Applied Physiology

Various explanations have been given for the haemostatic efficacy of the garment (Pelligra and Sandberg 1979). Application of circumferential pressure to a blood vessel decreases the radius of the vessel. Poiseuille's law states that laminar flow through a tube is directly proportional to the fourth power of the radius of the tube as determined by the equation:

$$Q = \frac{P\pi r^4}{8\eta l}$$

where Q = flow
   P = pressure gradient
   r = vessel radius
   η = viscosity
   l = length of vessel

Therefore external compression of the blood vessel will considerably reduce flow through it and decrease arterial bleeding.

The method by which inflation of the garment produces an increase in mean arterial blood pressure was originally thought to be due to an "auto-transfusion" effect providing an increased venous return and hence an increased cardiac output (Civetta et al. 1976). However, more recent studies in dogs and humans (Niemann et al. 1983; Abraham et al. 1984) found no significant increase in cardiac output. There was a significant correlation between the increase in mean aortic pressure and the increase in peripheral vascular resistance; this is thought to be the main mechanism by which PTs produce their effect in hypovolaemic shock.

Indications

The primary indication for the use of PTs is in the initial management of hypovolaemic shock, as a holding measure prior to volume resuscitation and surgery if necessary. Wayne and Macdonald (1983) reviewed their use in 1120 patients; the cause of shock in these patients included trauma, burns, major gastrointestinal, vascular and obstetrical haemorrhage, anaphylactic reactions and sepsis. Eight hundred and twenty-one (73%) of these patients survived longer than 24 hours and 756 patients were discharged alive. Fifty-eight per cent responded to PT application with an increase in blood pressure of 20 mmHg or to within normal limits; 17% responded with a fall in pulse rate to below 105 and 14% developed signs of improved tissue perfusion. In total 84% of patients showed one or more of these responses.

Precautions and Complications

The compartment syndrome has been reported after the use of PTs, particularly when applied over lower-extremity fractures (Mauli et al. 1981). This has also been reported in a trauma victim with uninjured legs (Williams et al. 1982), suggesting that a cause and effect relationship may exist. Other potential complications are ischaemic skin changes and poor renal perfusion; Wayne and Macdonald (1983) noted a 4% incidence of ischaemic skin changes (none of which required grafting) and a 0.97% incidence of renal failure requiring haemodialysis.

Respiratory function may be impaired when the abdominal compartment of the trousers is inflated; in clinical practice, however, this has not been a problem (Ransom and McSwain 1978; Wayne and Macdonald 1983).

Palafox et al. (1981) investigated the effect of inflation of PTs on intracranial pressure (ICP) in dogs. In the presence of an experimentally induced intracranial mass lesion, PT inflation in hypovolaemic dogs produced a rise in systolic blood pressure and also in central venous pressure (CVP). The ICP rose with the increase in CVP; however, cerebral perfusion pressure was also increased and the ICP did not rise to harmful levels. However, in the presence of a tension pneumothorax or cardiac tamponade the rise in CVP and ICP was much greater and did reach harmful levels. The presence of cardiac tamponade, intrathoracic bleeding or tension pneumothorax is a relative contraindication to the use of PTs (Davis et al. 1981).

Inflation of the abdominal compartment may lead to vomiting or regurgitation followed by aspiration in the comatose patient (Butson 1983). In such a situation the airway should therefore be protected. After deflation of the trousers there may be a fall in arterial pH. In one large study (Wayne and Macdonald 1983), however, mean pH before deflation was 7.36 and after deflation was 7.31. The acidosis was primarily controlled by ventilation.

Conclusion

The use of mechancial assistance devices should be considered early in the management of shock. The complications associated with the use of these techniques must be weighed against their ability to provide a temporary stabilization or improvement in the clinical situation; this may allow enough time for definitive treatment, either medical or surgical, to be instituted without the detrimental cardiorespiratory effects that inotropic support would have.

References

Abraham E, Cobo JC, Bland RD, Shoemaker WC (1984) Cardiorespiratory effects of pneumatic trousers in critically ill patients. Arch Surg 119:912–915

Alcan K, Stertzer SH, Wallish E, Bruno MS, Depasquale NP (1984) Current status of intra-aortic balloon counterpulsation in critical care cardiology. Crit Care Med 12:489–495

Baxter FJ, Kessaram RA, Baillie F (1984) Pneumatic antishock trousers. Can J Surg 27:423–427

Berger RL, Virender SK, Long W, Hechtman M, Hood V (1973) The use of diastolic augmentation with the intra-aortic balloon in human septic shock with associated coronary artery disease. Surgery 74:601–606

Braunwald E (1971) Control of myocardial oxygen consumption: Physiologic and clinical considerations. Am J Cardiol 27:416–432

Butson ARC (1983) The clinical use of antishock trousers. Can Med Assoc J 128:1428

Civetta JM, Nussenfeld SR, Rowe TR, Hirschman JC, McCullough KE, Nagel EL, Kaplan BH, Pettyjohn FS (1976) Pre-hospital use of the military antishock trouser (MAST). J Am Coll Emerg Phys 5:581–587

Clauss RH, Birtwell WC, Albertal G, Lunzer S, Taylor WJ, Fosberg AM, Harken DE (1961) Assisted circulation I. The arterial counterpulsator. J Thorac Cardiovasc Surg 41:447–458

Committee on Trauma (1977) Committee on Trauma, American College of Surgeons: essential equipment for ambulances. Bull Am Coll Surg 62:7–12

Cutler BS, Daggett WM (1971) Application of the "G-suit" to the control of hemorrhage in massive trauma. Ann Surg 173:511–514

Dannewitz SR, Lilja GP, Ruiz E (1981) Effect of pneumatic trousers on intracranial pressure in hypovolaemic dogs with an intracranial mass. Ann Emerg Med 10:176–181

Davis JW, McKone TK, Cram AE (1981) Hemodynamic effects of military antishock trousers (MAST) in experimental cardiac tamponade. Ann Emerg Med 10:185–186

Dunn JM, Kirsch MM, Harness J, Lee R, Straker J, Sloan H (1974) The role of assisted circulation in the management of endotoxic shock. Ann Thorac Surg 17:574–583

Ehrich DA, Biddle TL, Krowenberg MW, Yu PN (1977) The haemodynamic response to intra-aortic balloon counterpulsation in patients with cardiogenic shock complicating acute myocardial infarction. Am Heart J 93:274–280

Fuchs RM, Brin KP, Brinker JA, Guzman PA, Heuser RR, Yin FCP (1983) Augmentation of regional coronary blood flow by intra-aortic balloon counterpulsation in patients with unstable angina. Circulation 68(1):117–123

Goldberger M, Tabak SW, Prediman KS (1986) Clinical experience with intra-aortic balloon counterpulsation in 112 consecutive patients. Am Heart J 111:497–502

Goldman BS, Hill TJ, Rosenthal GA, Scully HE, Weisel RD, Baird RJ (1982) Complications associated with use of the intra-aortic balloon pump. Can J Surg 25:153–156

Hagemeijer F, Laird JD, Hallebos MMP, Hugenholtz PG (1977) Effectiveness of intra-aortic balloon pumping without cardiac surgery for patients with severe heart failure secondary to a recent myocardial infarction. Am J Cardiol 40:951–956

Hinshaw LB (1974) Role of the heart in the pathogenesis of endotoxic shock. A review of the clinical findings and observations on animal species. J Surg Res 17:134–145

Hinshaw LB, Shanbour LL, Greenfield LJ (1970) Mechanism of decreased venous return. Subhuman primate administered endotoxin. Arch Surg 100:600–606

Hoffman JR (1980) External counterpressure and the MAST suit: current and future roles. Ann Emerg Med 9:419–421

Isner JM, Cohen SR, Virmani R, Lawrinson W, Roberts WC (1980) Complications of the intra-aortic balloon counterpulsation device: clinical and morphologic observations in 45 necropsy patients. Am J Cardiol 45:260–268

Kantrowitz A, Tjonneland S, Freed PS, Phillips SJ, Butner AN, Sherman JL (1968) Initial clinical experience with intra-aortic balloon pumping in cardiogenic shock. JAMA 203:113–118

Leinbach RC, Buckley MJ, Austen WG, Petschek ME, Kantrowitz AR, Sanders CA (1971) Effects of intra-aortic balloon pumping on coronary flow and metabolism in man. Circulation 43(1):77–81

Leinbach RC, Gold HK, Dinsmore RE, Munoth ED, Buckley MJ, Austen WG, Sanders CA (1973) The role of angiography in cardiogenic shock. Circulation 48(3):95–98

Mauli KI, Capehart JE, Cardea JA, Haynes BW (1981) Limb loss following Military Anti-Shock Trousers (MAST) application. J Trauma 21:602

McEnany MT, Kay HR, Buckley MJ, Daggett WM, Erdmann AJ, Munoth ED, Rao RS, Detoeuf J, Austen WG (1978) Clinical experience with intra-aortic balloon pump support in 728 patients. Circulation 58(1):124–132

McSwain NE (1977) Pneumatic trousers and the management of shock. J Trauma 17:719–724

Mercer D, Doris P, Salerno TA (1981) Intra-aortic balloon counterpulsation in septic shock. Can J Surg 24:643–645

Michels R, Kint PP, Hagemeiser F, Baakumaran K, Van Der Brand M, Serruys PW, Hugenholtz PG (1980) Intra-aortic balloon pumping in myocardial infarction and unstable angina. Eur Heart J 1:31–34

Moulopoulos SD, Topaz S, Kolff WJ (1962) Diastolic balloon pumping (with carbon dioxide) in the aorta: a mechanical assistance to the failing circulation. Am Heart J 63:669–675

Niemann JT, Stapczynski JS, Rosborough JP, Rothstein RJ (1983) Haemodynamic effects of pneumatic external counterpressure in canine hemorrhage shock. Ann Emerg Med 661:13–16

Palafox BA, Johnson MN, McEwen DK, Gazzaniga AB (1981) ICP changes following application of the MAST suit. J Trauma 21:55–59

Pelligra R, Sandberg EC (1979) Control of intractable abdominal bleeding by external counterpressure. JAMA 241:708–713

Pennington DG, Swartz M, Codd JE, Merjavy JP, Kallser GC (1983) Intra-aortic balloon pumping in cardiac surgical patients: a nine year experience. Ann Thorac Surg 36:125–130

Powell WJ, Daggett WM, Magro AE, Bianco JA, Buckley MJ, Sanders CA, Kantrowitz AR, Austen WG (1970) Effects of intra-aortic balloon counterpulsation on cardiac performance oxygen consumption and coronary blood flow in dogs. Circ Res 26:753–764

Ransom K, McSwain NE (1978) Respiratory function following application of MAST trousers. J Am Coll Emerg Phys 7:297–299

Resnekov L (1977) Circulatory support and early cardiac surgery in the management of cardiogenic shock complicating myocardial infarction. Ann Clin Res 9:134–143

Roberts AJ, Hoover EL, Alonso DR, Combes JR, Dineen P, Gay WA, Subramanian VA (1979) Prolonged intra-aortic balloon pumping in *Klebsiella* induced hypodynamic shock: cardiopulmonary, haematologic metabolic and pathological observations. Ann Thorac Surg 28:73–86

Saarnof SJ, Braunwald E, Welch GH, Case RB, Stainsby WN, Macruz R (1958) Haemodynamic determinants of oxygen consumption of the heart with special reference to the time-tension index. Am J Physiol 192:148–156

Sanfelippo PM, Baker NH, Ewy HG, Moore PJ, Thomas JW, Bramos GJ, McVicker RF (1986) Experience with intra-aortic balloon counterpulsation. Ann Thorac Surg 41:36–41

Scheidt S, Wilner G, Mueller H, Summers D, Lesch M, Waff G, Krakauer J, Rubenfire M, Fleming P, Noon G, Oldham N, Killip T, Kantrowitz A (1973) Intra-aortic balloon counterpulsation in cardiogenic shock. N Engl J Med 288:979–984

Sponitz HM, Covell JW, Ross J, Braunwald E (1969) Left ventricular mechanics and oxygen consumption during arterial counterpulsation. Am J Physiol 217:1352–1358

Tobias MA (1981) Intra-aortic balloon pumps. Br J Hosp Med 26:542–548

Wayne MA, Macdonald SC (1983) Clinical evaluation of the antishock trouser: retrospective analysis of five years of experience. Ann Emerg Med 342:13–18

Weber KT, Ratsmin RA, Janicki JS, Rackley CE, Russell RD (1973) Left ventricular dysfunction following acute myocardial infarction. A clinicopathologic and haemodynamic profile of shock and failure. Am J Med 54:697–705

Weisel RD, Vito L, Dennis RC, Valeri CR, Hechtman HB (1977) Myocardial depression during sepsis. Am J Surg 133:512–521

Willerson JT, Curry GC, Watson JT, Lesmin SJ, Ecker RR, Mullins CB, Platt MR, Sugg WL (1975) Intra-aortic balloon counterpulsation in patients in cardiogenic shock, medically refractory left ventricular failure and/or recurrent ventricular tachycardia. Am J Med 58:183–191

Williams TM, Knopp R, Ellyson JH (1982) Compartment syndrome after antishock trouser use without lower extremity trauma. J Trauma 22:595–597

Section V
The Diagnosis and Prognosis of the Adult Respiratory Distress Syndrome

Chapter 16

Prognosis in the Intensive Care Unit: General Principles and Application to Patients with the Adult Respiratory Distress Syndrome

J. R. Le Gall

There are many studies of outcome following an episode of shock or following the development of Adult Respiratory Distress Syndrome (ARDS) in patients requiring intensive therapy. This chapter will attempt to demonstrate the main principles of prognostic studies and give some results in these groups of patients.

General Principles of Prognostic Studies

In general, prognosis directly depends on four items: the previous health status of the patient, his or her age, the severity of his or her illness and the diagnosis. Moreover, some authors have proposed the use of the therapeutic effort as a prognostic indicator.

Previous Health Status

The system most usually employed in intensive care units (ICUs) is the chronic health status part of the APACHE (Acute Physiologic And Chronic Health Evaluation) published by W. A. Knaus and others in 1981 (Knaus et al. 1981). The existing health status 3 months before hospitalization is categorized into four groups: A, good health status, without any limitation; B, moderate limitation of activity; C, severe limitation; D, bedridden or institutionalized.

The previous health status may also be defined as pre-existing or no pre-existing chronic disease, as in the protocol study of ARDS published by the European Society of Intensive Care Medicine (Artigas et al. 1985). Chronic organ insuffiency before the hospital admission is defined according to the following criteria:

1. *Liver:* Biopsy-proven cirrhosis and documented portal hypertension or previous episodes of upper gastrointestinal bleeding attributed to portal hypertension, or prior episodes of hepatic failure encephalopathy/coma.
2. *Cardiovascular:* New York Heart Association Class IV, i.e. angina or symptoms at rest or during minimal exertion (e.g. getting dressed or self-care).
3. *Respiratory:* Chronic restrictive, obstructive or vascular diseases resulting in severe exercise restriction, i.e. unable to climb stairs or perform household duties; or documented chronic hypoxia, hypercapnia, secondary polycytaemia, severe pulmonary hypertension (mean pulmonary artery pressure >40 mmHg), or respiratory dependency.
4. *Renal:* Receiving chronic haemodialysis or peritoneal dialysis.
5. *Immunocompromised state:*

   (a) The patient has received immunosuppression therapy, chemotherapy, radiation, long-term low-dose steroids e.g. hydrocortisone (for at least 30 days prior to hospitalization) or recent high-dose steroids (15 mg kg^{-1} for 5 or more days),
       or
   (b) The patient has a disease that is sufficiently advanced to suppress resistance to infection (e.g. leukaemia, lymphoma, AIDS, documented diffuse metastatic cancer).

Age

The influence of age on prognosis is predominant. In a prospective study Le Gall and co-workers showed that prognosis was worse in old patients, as regards both survival rate and post-ICU health status (Le Gall et al. 1982). Nicolas, Le Gall and others found similar results in a multi-centre study (Nicolas et al. 1987). Thus age is often included in severity scores (Le Gall et al. 1983a, 1984; Knaus et al. 1985a).

Severity of Illness

Severity of illness may be appreciated by general or specialized scores. We shall describe later some specialized scores for ARDS. The two main types of general scores are severity indexes and organ system failure.

Severity Indexes

Many different severity scores have been proposed for use with ICU patients, and have been specifically applied to the assessment of burned, coronary care and trauma patients. The APACHE system has proved its reliability and validity in both national and international studies (Le Gall et al. 1983b; Knaus et al. 1982). Its physiological portion (APS) measures the degree of acute illness. It does this by surveying the clinical record within the first 24 hours of admission for abnormalities among 34 possible physiological measurements. A weight ranging from 0 to 4 is assigned to each recorded measurement. The sum of the assigned weights for all measurements recorded represents the patient's total physiological score. The higher the score, the sicker the patient.

Although APS is generally accepted as a reliable estimate of severity of illness in individual patients, variations in the mean number of data collected per patient may introduce a systematic bias in patient scoring because missing values are interpreted as normal. A Simplified Acute Physiologic Score (SAPS) was proposed by Le Gall and co-workers in 1983 (Le Gall et al. 1983a). It consists of 14 easily measured biological and clinical variables: 12 from the original APS, with the same weights (0 to 4); age of patient, with an assigned range of 0 to 4; and a fixed value of 3 assigned to ventilated patients (Table 16.1). SAPS compared favourably with the APS (Le Gall et al. 1984) and is simply a less time-consuming (1 or 2 minutes) method for comparative studies and management evaluation between different ICUs. SAPS is now routinely used in most French ICUs, and in those of several other European countries.

The APACHE II has been published by Knaus et al. (1985a). Ninety per cent of the selected items are identical to those in SAPS. The main difference consists of two additional variables: previous health status and diagnosis. Given that these two factors may have some influence on outcome, any improvement in reliability of prediction of outcome is moderate when physicians attribute a diagnosis and previous health status to their patients.

Acute Organ System Failure (OSF)

The OSF system was proposed by Knaus in 1985 (Knaus et al. 1985b). The definitions of OSF are given in Table 16.2. In 13 US hospitals, 5677 ICU admissions were monitored. The number and duration of OSF were linked to outcome at hospital discharge for each of the 2719 ICU patients (48%) who developed OSF. For all medical and most surgical admissions a single OSF lasting for more than 1 day resulted in a mortality rate approaching 40%; two OSF for more than 1 day increased death rates to 60%. Advanced chronological age increased both the probability of developing OSF and the probability of death once OSF occurred. Mortality for 99 patients with three or more OSF existing after 3 days was 98%.

Diagnosis

The diagnosis plays a key role in hospital outcome. For the same SAPS score, mortality is much higher for cardiogenic shock than for barbiturate overdose (Fig. 16.1). Nevertheless it is often difficult to define a precise or single diagnosis for an ICU patient. In addition, different physicians may give different diagnoses for the same patient. Predefined diagnosis is reliable; for example, acute pancreatitis may be defined by clinical, biological and CT scan criteria. On the other hand to attribute a precise or single diagnosis to each ICU patient is simply not realistic.

Table 16.1. Scoring values for the 14 variables of SAPS

Variable SAPS scale	4	3	2	1	0	1	2	3	4
Age (yr)					≤45	46–55	56–65	66–75	>75
Heart rate (beats min⁻¹)	≥180	140–179	110–139		70–109		55–69	40–54	<40
Systolic blood pressure (mmHg)	≥190		150–189		80–149		55–79		<55
Body temperature (°C)	≥41	39.0–40.9		38.5–38.9	36.0–38.4	34.0–35.9	32.0–33.9	30.0–31.9	<30.0
Spontaneous respiratory rate (breath/min)	≥50	35–49		25–34	12–24	10–11	6–9		<6
or									
Ventilation or CPAP								Yes	
Urinary output (l (24 h)⁻¹)			>5.00	3.50–4.99	0.70–3.49		0.50–0.69	0.20–0.49	<0.20
Blood urea (mmol l⁻¹)	≥55.0	36.0–54.9	29.0–35.9	7.5–28.9	3.5–7.4	<3.5			
Haematocrit (%)	≥60.0		50.0–59.9	46.0–49.9	30.0–45.9		20.0–29.9		<20.0
White blood cell count (10³ mm⁻³)	≥40.0		20.0–39.9	15.0–19.9	3.0–14.9		1.0–2.9		<1.0
Serum glucose (mmol l⁻¹)	≥44.5	27.8–44.4		14.0–27.7	3.9–13.9		2.8–3.8	1.6–2.7	<1.6
Serum potassium (mequiv l⁻¹)	≥7.0	6.0–6.9		5.5–5.9	3.5–5.4	3.0–3.4	2.5–2.9		<2.5
Serum sodium (mequiv l⁻)	≥180	161–179	156–160	151–155	130–150		120–129	110–119	<110
Serum bicarbonate (mequiv l⁻¹)		>40.0		30.0–39.9	20.0–29.9	10.0–19.9		5.0–9.9	<5.0
Glasgow coma score					13–15	10–12	7–9	4–6	3

Table 16.2. Definitions of organ system failure (OSF)

If the patient had one or more of the following during a 24 hour period (regardless of other values), OSF existed on that day:

1. Cardiovascular failure (presence of *one or more* of the following):
   (a) Heart rate $\leq$ 54 min^{-1}
   (b) Mean arterial blood pressure $\leq$49 mmHg
   (c) Ventricular tachycardia and/or ventricular fibrillation
   (d) Serum pH $\leq$7.24 with a PaCO$_2$ of $\leq$49 mmHg

2. Respiratory failure (presence of *one or more* of the following):
   (a) Respiratory rate $\leq$ 5 min^{-1} or >49min^{-1}
   (b) PaCO^2 $\geq$50 mmHg
   (c) D(A—a)O$_2$$\geq$350 mmHg (D(A–a)O$_2$ = 713 FiO$_2$ – PaCO$_2$ – PaO$_2$)
   (d) Dependent on ventilator on the fourth day of OSF, i.e. not applicable for the initial 72 h of OSF

3. Renal failure (presence of *one or more* of the following):[a]
   (a) Urine output $\leq$479 ml (24 h)$^{-1}$ or $\leq$159 ml (8 h)$^{-1}$
   (b) Serum blood urea nitrogen $\geq$100 mg 100 ml^{-1}
   (c) Serum creatinine $\geq$3.5 mg 100 ml^{-1}

4. Haematological failure (presence of *one or more* of the following):
   (a) White blood cell count $\leq$1000 mm^3
   (b) Platelets $\leq$ 20 000 mm^3
   (c) Haematocrit $\leq$ 20%

5. Neurological failure (Glasgow Coma Score <6)

[a]Excluding patients on chronic dialysis before hospital admission.

Abbreviations: PaCO$_2$, arterial carbon dioxide tension; D(A–a)O$_2$, arterial–alveolar oxygen gradient; FiO$_2$, fraction of inspired oxygen; PaCO$_2$, arterial carbon dioxide tension.

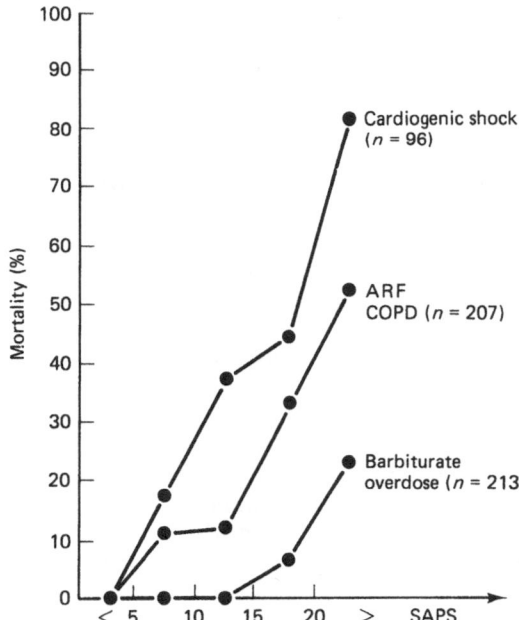

Fig. 16.1. Influence of diagnosis on outcome (from the French study of 3688 patients). ARF, acute respiratory failure; COPD, chronic obstructive pulmonary disease.

Intensity of Therapy

The Therapeutic Intervention Scoring System (TISS), proposed by Cullen in 1974 (Cullen et al. 1974) and updated in 1983 (Keene and Cullen 1983), has become a widely accepted method for classifying critical care patients according to their prognosis. Patients were classified by Cullen in four groups: class I, routine recovery after uneventful anaesthesia and operation; class II, close observation for potential catastrophe in a patient whose condition was otherwise stable; class III, intensive care nursing for a heavily monitored, stabilized patient; and class IV, intensive physician and nursing care for a patient whose condition was unstable and whose clinical course was unpredictable, requiring frequent changes of orders and therapy. The average stay in the ICU was from 4 hours to 2.7 days and the total of TISS points for the first 24 hours varied from 5 in the class I patients to 43 in class IV. Mortality rate at 1 year was 0% in class I, 73% in class IV. Thus, the TISS system can be used to classify patients: class IV, more than 40 points; class III, 20–39 points; class II, 10–19 points; class I, less than 10 points.

Outcome

Outcome is usually noted as the mortality rate or survival rate. The ICU outcome is not a very good indicator for comparing ICUs or comparing patients suffering from the same disease because the different policies from one hospital to another result in different lengths of stay in ICU. It is more valuable to use the hospital outcome, which is usually taken as the survival rate at 2 or 3 weeks. The best measure of outcome is the survival rate at 1 year or 6 months (Le Gall et al. 1982), but data collection is often difficult.

The health status of the survivors is rarely mentioned. Nevertheless in a study of 112 survivors after 1 year, Le Gall et al. (1982) showed that 62% had the same health status before and after ICU treatment; other than one patient, whose health status had improved, the remaining patients (37%) had a worsened health status 1 year later.

Is Each Item Independent of the Other Ones?

Processes depend on the input, and the outcome depends on both processes and input. It is important to remember, however, that the elements of input are not independent of each other.

Fig. 16.2 shows the logical relationship between these parameters. Each of the four input points is correlated with the three others. For instance, previous health status is statistically worse when age increases (Le Gall et al. 1982); severity of illness is higher when previous health status is worse (Knaus et al. 1981); some diagnoses are more frequent in young or in old people. In the same way, each of the four input points is correlated with TISS and outcome. For instance, age plays a role in the length of stay in hospital and in the number of TISS points (Nicolas et al. 1987). The role of age on outcome is less marked when severity of illness increases (Nicolas et al. 1987). Diagnosis markedly influences outcome: SAPS score is correlated with mortality in diagnosis-related groups of patients, but for the same SAPS score outcome is very different according to the diagnosis (Le Gall et al. 1983b). TISS points have also been

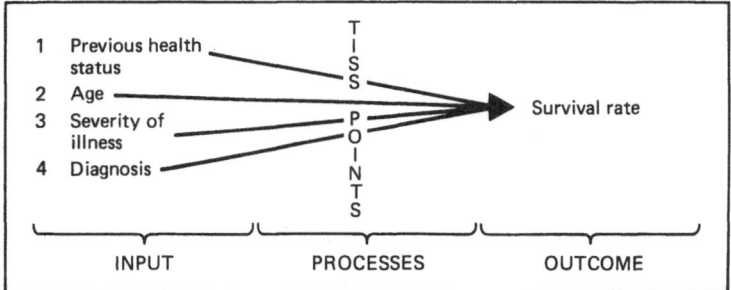

Fig. 16.2. Relationships between descriptive items of ICUs.

correlated with outcome, as demonstrated by Cullen et al., since they were able to define groups with increasing severity of illness (from class I to class IV), with a mortality rate at 1 year varying from 0 to 73%, according to the TISS point score on the first day of ICU.

Prognosis of ARDS

Prognostic indices used in ARDS studies can be either specific or general.

Specific Indices

Jardin et al. (1982) have proposed an *oxygenation index* calculated from arterial oxygenation:

$$P(A-a)O_2/PAO_2 + 0.014 \text{ PEEP (cm } H_2O)$$

Where $P(A-a)O_2$ is alveolar–arterial oxygen difference, PAO_2 is alveolar oxygen tension and PEEP is positive end-expiratory pressure. This index was used from the second day of treatment onward to predict with a strong probability whether the lung lesions would subside or develop in severity.

Morel et al. (1985) have described a *pulmonary failure scoring system* with five severity degrees for each of four variables (one radiographic and three physiological) (Table 16.3). In fact, this score is more descriptive than predictive: all patients with ARDS have a score of 2 or more. However, the same group of investigators have proposed some other very specific markers of ARDS and its severity, such as the pulmonary extraction of serotonin.

General Indices

The APACHE system has been used by Montgomery et al. (1985) in non-trauma patients, while the Injury Severity Score (ISS: Baker et al. 1974) was used in trauma

Table 16.3. Pulmonary failure scoring system

Score	Chest roentgenogram	$P(A-a)O_2/FiO_2$ (mmHg)[a]	C_{rs} (ml $(cm\,H_2O)^{-1}$)	$P\bar{A}P$ (mmHg)
0	Normal	<300	>80	<20
1	Moderately increased interstitial marking	300–375	70–80	20–25
2	Markedly increased interstitial marking	375–450	50–70	25–30
3	Patchy air-space consolidation	450–525	30–50	30–35
4	Extensive air-space consolidation	>525	<30	>35

Abbreviations: $P(A-a)O_2/FiO_2$, alveolar–arterial oxygen tension difference divided by inspired oxygen fraction; C_{rs}, static compliance of the respiratory system; $P\bar{A}P$, mean pulmonary arterial pressure.
[a]Continuous positive pressure ventilation with continuous positive airway pressure (CPAP) or positive end-expiratory pressure (PEEP) of more than 5 cm H_2O increases this particular score by +1.

patients. They studied prospectively 207 patients identified as being at risk for development of ARDS. Forty-seven patients developed ARDS and the remaining 160 were used as a comparison control group. In both groups a linear relation was observed between APACHE score and the transformed probability of death. For any given APACHE score, patients with ARDS were more likely to die than those without ARDS. In trauma patients there was a linear relation between the ISS and transformed probability of death for patients without ARDS, but no particular relation for those with ARDS. In this study only 16% of deaths in the ARDS group were from irreversible respiratory failure. This might explain why non-specific indices such as APACHE are reliable in ARDS. This syndrome frequently results in multi-organ system failure via septic mechanisms, and this is usually the predominant factor in a fatal outcome.

Another interesting study was that of Fowler et al. (1985) who, using discriminant analysis, showed that the percentage of band forms in the differential white cell count, together with the arterial pH and bicarbonate level during mechanical ventilation, were variables significantly associated with survival. The relative risk was 1.0 when there were no excess-risk variables present; this rose to 2.0–2.7 with one risk variable, 4.0–5.4 with two variables, and to 10.6 when the three risk variables were present together.

Table 16.4. Evolution of a general (SAPS) and a specific (Jardin's JI) index in surviving (S) patients compared with those who subsequently die (non-S) with ARDS (from Artigas et al. 1985)

Days:		1	3–5	>5
SAPS	S	12.6 ± 4.3	9.9 ± 3.8	9.2 ± 3.3
	non-S	14.3 ± 4.1	14.7 ± 5.2	***16.4 ± 5.1
JI	S	0.8 ± 0.09	*0.5 ± 0.11	**0.47 ± 0.01
	non-S	0.8 ± 0.08	0.7 ± 0.13	0.6 ± 0.014

*$p < 0.025$; **$p < 0.005$; ***$p < 0.001$.
Both prognostic indices are better able to predict death after the fifth day. On days 3–5 only SAPS was predictive of outcome.

Artigas and Mancebo (1985) compared Jardin's oxygenation index with SAPS in 35 patients with severe ARDS. Three controls were performed at 1 day, 3–5 days and at more than 5 days. Total mortality was 69%. The results are shown in Table 16.4. Patients who subsequently died did not demonstrate any improvement in their severity indices during the course of the illness. In contrast, both SAPS and the JI improved significantly during the clinical evolution of survivors.

All these studies point to one important conclusion: general indices are useful as prognostic indicators in ARDS because death is often related to an extrapulmonary disturbance in organ function. ARDS is not just a form of acute lung disease, but more often is associated with a widespread systemic disturbance.

References

Artigas A, Mancebo J (1985) Comparative study of different severity indexes in acute respiratory failure. In: 4th world congress on intensive and critical care medicine. King and Wirth, London, p 90

Artigas A, Carlet J, Chastang C, Le Gall JR, Cox P (1985) Protocol of adult respiratory distress syndrome study: clinical predictors, prognosis factors and outcome. European Society of Critical Care Medicine, Barcelona

Baker SP, O'Neill B, Hadden W, Long WB (1974) The injury severity score: a method for describing patients with multiple injuries and evaluating emergency care. J Trauma 14:187–196

Cullen DJ, Civetta JM, Briggs BA et al. (1974) Therapeutic intervention scoring system: a method for quantitative comparison of patient care. Crit Care Med 2:57–63

Fowler AA, Hamman F, Zerbe GO, Benson KN, Hyers TM (1985) Adult respiratory distress syndrome: prognosis after onset. Am Rev Respir Dis 132:472–478

Jardin F, Prost JF, Bazin M et al. (1982) Modalités évolutives du syndrome de détresse respiratoire aiguë de l'adulte. Nouv Presse Méd 11:29–33

Keene AB, Cullen DJ (1983) Therapeutic Intervention Scoring System: update 1983. Crit Care Med 11:1–3

Knaus WA, Zimmerman JE, Wagner DR et al. (1981) APACHE : Acute Physiology And Chronic Health Evaluation: a physiologically based classification system. Crit Care Med 9:591–597

Knaus WA, Le Gall JR, Wagner D et al. (1982) A comparison of intensive care in the USA and France. Lancet II:642–646

Knaus WA, Draper EA, Wagner DP, Zimmerman JE (1985a) APACHE II: a severity of disease classification system. Crit Care Med 13:818–829

Knaus WA, Draper EA, Wagner D, Zimmerman JE (1985b) Prognosis in acute organ system failure. Ann Surg 6:685–693

Le Gall JR, Brun-Buisson C, Trunet P et al. (1982) Influence of age, previous health status and severity of illness on outcome from intensive care. Crit Care Med 10:575–577

Le Gall JR, Loirat P, Alperovitch A (1983a) Simplified Acute Physiological Score for intensive care patients. Lancet II:741

Le Gall JR, Loirat P, Nicolas F et al. (1983b) Utilisation d'un indice de gravité dans huit services de réanimation multidisciplinaire. Press Méd 12:1757–1761

Le Gall JR, Loirat P, Alperovitch A (1984) A Simplified Acute Physiology Score for ICU patients. Crit Care Med 12:975–977

Montgomery AB, Stager MA, Carrico LJ, Hudson LD (1985) Causes of mortality in patients with the ARDS. Am Rev Respir Dis 132:485–499

Morel D, Dargent F, Bachmann M, Suter PM, Junot AF (1985) Pulmonary extraction of serotonin and propranolol in patients with ARDS. Am Rev Respir Dis 132:479–484

Nicolas F, Le Gall JR, Alperovitch A, Loirat P, Villers D (1987) Influence of patient's age on survival, level of therapy and length of stay in intensive care units. Intensive Care Med (in press)

Chapter 17

Adult Respiratory Distress Syndrome: A Scoring System for the Estimation of the Gravity of Pulmonary Disease and Comparison of Patient Populations

P. M. Suter

The adult respiratory distress syndrome (ARDS) still carries a high mortality despite remarkable progress in our understanding of the pathophysiology and the introduction of new types of pharmacological and ventilatory support. However the figures for both incidence and mortality vary widely in the literature. This may in part be due to different definitions used in acute respiratory failure. Similarly, the efficiency of preventive and therapeutic actions are judged very differently in seemingly similar patient populations, because no uniform criteria for the severity of ARDS have emerged.

It is surprising to note that since the original description of this syndrome by Asbaugh et al. in 1967, the definition of ARDS has remained essentially unchanged (Petty and Fowler 1982). Some important pathophysiological changes associated with ARDS that have been recognized during the last 10 years, such as pulmonary hypertension (Zapol and Snider 1977), aggregation of polymorphonuclear leucocytes in the pulmonary capillaries and the activation of the complement pathway (Jacob et al. 1980), the coagulation cascade (Carvalho et al. 1982) and production of arachidonic acid metabolites (Brigham 1982), have not been used either quantitatively or qualitatively to characterize the severity of pulmonary parenchymal disease.

The purpose of the present chapter is to review briefly some classifications of ARDS described according to its severity, and to propose a simple scoring system of acute pulmonary parenchymal failure taking into account four independent variables which are currently available in these patients in the intensive care unit.

Indexes of Severity Reported

During the last few years a number of investigators in the United States and Europe (Shapiro et al. 1983; W. Zapol, pers. comm. 1985) have classified patients at risk for ARDS or with ARDS using the categories:

    patients at risk
    mild ARDS
    moderate ARDS
    severe ARDS

These levels of severity were obtained on the basis of the medical history containing one or more risk factors known to be associated with a certain probability of developing ARDS—such as trauma, shock, sepsis or similar acute events (Fowler et al. 1983). The following three clinical elements were also taken into account:

    hypoxaemia
    requirement for some form of respiratory therapy
    changes in the chest radiograph

Table 17.1.   Classification and outcome of 334 patients studied by the Massachusetts General Hospital Acute Respiratory Failure score between December 1978 and July 1982 (W. Zapol, pers. comm. 1985)

Highest classification	No. of patients	Mean age, range (years)	Percentage mortality
At risk	59	45 (16–86)	7[a]
Mild	144	49 (17–86)	17[a]
Moderate	91	56 (16–87)	51
Severe	40	42 (16–76)	86
Total	334	50 (16–87)	

[a] Non-pulmonary-related death.

This type of categorization (W. Zapol, pers. comm. 1985) has allowed some investigators to distinguish patient groups with different rates of mortality (Table 17.1). The criteria used, however, do not include variables of respiratory function except arterial oxygen tension (PaO_2), nor measurements of pulmonary haemodynamics. Some authors have carried this type of classification one step further (Shapiro et al. 1983), suggesting a correlation of the stages cited with the morphological alterations in the lung parenchyma. Thereby, mild or moderate stages of acute lung injury were associated with dysfunction of the capillary endothelial cells and permeability, while alterations of alveolar epithelium were proposed to be present in severe forms of ARDS (Table 17.2).

  Addressing a somewhat different question, Fowler et al. identified three risk factors which, when present on the day of onset of ARDS, were associated with a sig-

Table 17.2. Clinical spectrum of acute lung injury (ALI) (from Shapiro et al. 1983)

	Endothelial cell dysfunction, increased permeability		Epithelial cell dysfunction, diminished surfactant function
Spectrum of ALI	Mild	Moderate	Severe
Clinical diagnosis[a]	NCE		ARDS
Degree of hypoxaemia	Moderate		Severe
Response fraction of inspired oxygen	Responsive		Refractory
Lung compliance	Moderate decrease		Severe decrease

From Shapiro et al. (1983).
[a]NCE, non-cardiogenic oedema; ARDS, adult respiratory distress syndrome.

nificantly higher mortality rate (Fowler et al. 1985). The three predictors of an increased mortality were:

1. An arterial pH below 7.40 (on the ventilator)
2. A total bicarbonate level in the serum below 20 mmol dl^{-1}
3. The presence of less than 10% band forms in the differential leucocyte count of the peripheral blood smear.

Mortality in patients with ARDS was reported to be about 40% when none of these risk factors was present on the day of onset, about 70% with one risk factor, 90% with two, and 100% when all three factors were present.

Investigating the different causes of death in patients with ARDS, Montgomery et al. (1985) observed that only 16% of the deaths were directly due to irreversible respiratory failure, the majority of the patients who died within the first 3 days after onset of ARDS doing so from the underlying injury or illness. The late deaths were primarily caused by sepsis, with the abdomen being the predominant source.

The "Geneva" ARDS Scoring System

In order to analyse the severity of ARDS as precisely as possible with methods available at the bedside in every intensive care unit, a simple scoring system was developed (Morel et al. 1985). This analysis uses four independent variables, each of which is assigned a score between 0 and 4. A score of 0 corresponds to a normal state, 4 the most diseased state. The four variables are (Table 17.3):

1. *Chest roentgenogram*, evaluated and scored by four experienced radiologists independently, without knowing the patient's history or clinical status.
2. *Alveolar–arterial oxygen tension difference*, corrected for the various inspired oxygen concentrations (FiO$_2$) administered by dividing by the FiO$_2$, and adapted for the effect of the application of positive airway pressure, because this treatment usually increases arterial oxygenation in this type of patient.

Table 17.3.   Geneva ARDS scoring system

Score	Chest roentgenogram	$D(A-a)O_2/FiO_2$ (mmHg)[a]	C_{rs} (ml (cm H_2O)$^{-1}$)	$P\bar{A}P$ (mmHg)
0	Normal	<300	>80	<20
1	Moderately increased interstitial marking	300–375	70–80	20–25
2	Markedly increased interstitial marking	375–450	50–70	25–30
3	Patchy air-space consolidation	450–525	30–50	30–35
4	Extensive air-space consolidation	>525	<30	>35

Abbreviations: $D(A-a)O_2/FiO_2$, alveolar–arterial oxygen tension difference divided by inspired oxygen fraction; C_{rs}, static compliance of the respiratory system; $P\bar{A}P$, mean pulmonary arterial pressure.
[a]Continuous positive pressure ventilation with continuous positive airway pressure (CPAP) or positive end-expiratory pressure (PEEP) of more than 5 cm H_2O increases this particular score by +1.

3. *Total static respiratory compliance* (C_{rs}), calculated in intubated patients by inflating the lungs with a graduated 1l-syringe from functional residual capacity, at atmospheric pressure, to a volume of 12 ml kg^{-1} body weight, and by recording airway pressure after a 2–4 second inspiratory pause. These measurements can also be obtained with the patient on the ventilator, using a slow inspiratory flow and an end-inspiratory hold of 2 seconds or longer. C_{rs} is calculated by dividing the volume change by the simultaneous airway pressure difference.

4. *Pulmonary artery pressure*, obtained from a Swan–Ganz catheter, the mean value being calculated electronically.

By dividing the total number of points scored by the number of factors taken into consideration, a mean score is obtained for each patient.

Validation of this scoring system was performed in two ways. First, we found a significant correlation between the mean score and a well-known metabolic function of the pulmonary endothelial cell, i.e. serotonin uptake (Junod 1975; Morel et al. 1985). Second, the score was related to the level of ventilatory support required and outcome (Morel et al. 1985). However, an evaluation in a larger patient population is necessary. Such an attempt is actually in progress in the context of the European Multicenter ARDS Study (A. Artigas, pers. comm. 1985).

In conclusion, the proposed ARDS scoring system should allow a more precise definition of the gravity of pulmonary dysfunction, and thereby make possible the comparison of patient populations and the efficiency of therapeutic interventions (Petty 1985).

References

Ashbaugh DG, Bigelow DB, Petty TL, Levine BE (1967) Acute respiratory distress in adults. Lancet II:319–323
Brigham KL (1982) Mechanisms of lung injury. Clin Chest Med 3:9–24
Carvalho AC, Bellman SM, Saullo VJ, Quinn DB, Zapol WM (1982) Altered factor VII in acute respiratory failure. N Engl J Med 307:1113–1119

Fowler AA, Hamman RF, Good JT, Benson KN, Baird M, Eberle DJ, Petty TL, Hyers TM (1983) Adult respiratory distress syndrome: risk with common predispositions. Ann Intern Med 98:593–597

Fowler AA, Hamman RF, Zerbe GO, Benson KN, Hyers TM (1985) Adult respiratory distress syndrome. Prognosis after onset. Am Rev Respir Dis 132:472–478

Jacob HS, Craddock PR, Hammerschmidt DR, Moldow CF (1980) Complement-induced granulocyte aggregation: an unsuspected mechanism of disease. N Engl J Med 302:789–794

Junod AF (1975) Metabolism, production and release of hormones and mediators in the lung. Am Rev Respir Dis 112:93–108

Montgomery AB, Stager MA, Carrico CJ, Hudson LD (1985) Causes of mortality in patients with the adult respiratory distress syndrome. Am Rev Respir Dis 132:485–489

Morel DR, Dargent F, Bachmann M, Suter PM, Junod AF (1985) Pulmonary extraction of serotonin and propranolol in patients with adult respiratory distress syndrome. Am Rev Respir Dis 132:479–484

Petty TL (1985) Indicators of risk, course, and prognosis in adult respiratory distress syndrome (ARDS) (editorial). Am Rev Respir Dis 132:471

Petty TL, Fowler AA III (1982) Another look at ARDS. Chest 82:98–104

Shapiro BA, Cane RD, Harrison RA (1983) Positive end-expiratory pressure in acute lung injury. Chest 83:558–563

Zapol W, Snider MT (1977) Pulmonary hypertension in severe acute respiratory failure. N Engl J Med 296:476–480

Chapter 18

The Clinical Presentation and Diagnosis of the Adult Respiratory Distress Syndrome

A. Lawson and D. Bihari

Patients requiring intensive care frequently develop arterial hypoxaemia and few clinicians doubt the importance of this particular complication and its management in the overall prognosis of the critically ill. Acute respiratory failure may occur following a wide variety of different insults which are traditionally divided into two groups—those which produce direct mechanical or chemical trauma to the lung (contusion, pneumonitis following exposure to noxious gases or subsequent to the aspiration of gastric contents) and those in which the lung damage is a consequence of some distant but systemic disease process (e.g. the sepsis syndrome, severe trauma, pancreatitis). The severity of pulmonary dysfunction is extremely variable. Indeed, only a small proportion of the hypoxaemic population, resistant to the effects of oxygen or continuous positive airway pressure by facemask, require endotracheal intubation and mechanical ventilation. Similarly, of those patients requiring assisted ventilation, only a small minority go on to demonstrate the classical features of the so-called adult respiratory distress syndrome (ARDS).

For these reasons, difficulties have arisen in the use of the term ARDS, which has now replaced the more descriptive "wet lung", "shock lung" and "Da Nang lung". Moreover, the term has been widely applied to include many different forms of acute respiratory failure with very different prognoses. In particular, varying definitions of the syndrome have led to some controversy concerning the changing mortality rate of acute respiratory failure in the intensive care unit (ICU) and the interpretation or relevance of a number of clinical trials (Newman 1983; Andreadis and Petty 1985; anonymous 1986a, b).

Definitions

Most investigators use the term ARDS to describe that form of acute respiratory failure which is characterized by severe refractory hypoxaemia, a falling pulmonary compliance with an increasing shunt fraction in the absence of a raised pulmonary capillary pressure, leading to dyspnoea, tachypnoea and cyanosis in the spontaneously breathing patient. The chest X-ray appearances of diffuse bilateral pulmonary infiltrates are usually included in the clinical definition of the syndrome but are not essential for its diagnosis especially in patients receiving continuous positive airway pressure (CPAP) or positive end-expiratory pressure (PEEP). Although seemingly precise, this description includes many cases of severe non-lobar bacterial or viral pneumonia and the inclusion of such patients in studies of the natural history of ARDS has contributed in part to the confusion surrounding overall survival. As a part of the spectrum of acute respiratory failure and depending upon the criteria used to select patients with ARDS (What is severe refractory hypoxaemia? On what concentration of oxygen? With or without CPAP or PEEP? On or off the ventilator? In the presence or absence of a normal plasma colloid oncotic pressure? In the presence or absence of a primary pneumonia? In the presence or absence of chronic lung disease?), it is apparent that there are mild, moderate and severe forms of the disease (Pontoppidan et al. 1972, 1985).

The description of ARDS by Ashbaugh and Petty in 1967 was fundamental in the definition of the syndrome but was not original in so far as many previous authors had identified abnormalities in lung structure and function as one sequel of an episode of shock. Almost 40 years ago, Moon (1948) reviewed 112 cases of secondary shock at post-mortem and found that pulmonary atelectasis occurred in at least 50% of cases studied while oedema and congestion was present in all. Even then, changes in pulmonary capillary permeability were thought to be an essential feature of the lung pathology of secondary shock and this increased permeability has remained an essential characteristic in the pathophysiological description of the clinical condition (Robin et al. 1972). Nevertheless, it was left to Ashbaugh and Petty to group together patients with severe acute respiratory failure of different aetiologies and to emphasize that the clinical response of the acutely injured lung was and is strictly limited. Furthermore, this original description of the syndrome included some extremely pertinent comments on the use of PEEP and corticosteroids in the management of the associated refractory hypoxaemia, the accuracy of which has survived the passing of two decades.

At the hub of the problem is survival. Ashbaugh and Petty reported 5 survivors amongst their 12 patients giving a mortality rate of 60%, and it is frequently stated that this has not changed over the last 20 years despite the introduction of highly sophisticated support systems (anonymous 1986a, b; Pepe 1986). This is not our view. We believe it is yet another example of fundamental changes in the spectrum and severity of disease, so that the modern ICU attempts to treat patients with difficult, if not irreversible underlying primary problems—patients experiencing massive trauma, severe burns, prolonged and extensive surgery who previously would not have survived to reach the ICU; patients with malignancies, neutropaenia, acquired immunodeficiency syndromes, following bone marrow transplantation etc.—in whom the management of acute respiratory failure is just one small aspect of their care. Similar concern surrounds the unchanging mortality rate of 50%–80% from acute renal failure in the ICU (Wardle 1982; Cameron 1986), but it is interesting to note that a very similar group of patients is in fact being considered (Wheeler et al. 1986). Not surprisingly, the development of severe combined acute renal and respiratory failure (with the acronym SCARRF) in the immunocompromised cancer patient with sepsis is bad news, but it emphasizes the importance of other organ failure in the eventual outcome from critical illness originating with either acute respiratory or renal failure as the primary event.

A substantial number of studies have now confirmed that the primary cause of death in patients with so-called ARDS is not the inability to oxygenate arterial blood adequately but rather the result of the development of multiple organ dysfunction and failure (Petty 1985; Montgomery et al. 1985; Fowler et al. 1985; Hyers and Fowler 1986). In recent investigations as few as 15% of deaths have been "pulmonary" in origin related to intractable respiratory failure, whereas 75% have appeared to be more closely associated with the development of the sepsis syndrome and other organ dysfunction (Bartlet et al. 1986; Seidenfeld et al. 1986). Infection, primary or nosocomial, seems to be an essential factor in triggering other organ dysfunction and Johanson's group reported a prevalence of multiple organ systems dysfunction of 93% in infected patients with ARDS compared with only 47% in supposedly uninfected cases (Bell et al. 1983). Most physicians now accept that ARDS is not purely a respiratory disease but more a reflection of a disseminated disorder following on from or coinciding with a period of respiratory failure and producing widespread tissue damage (anonymous 1986a, b; Rinaldo and Rogers 1986).

More emphasis has now been placed upon defining populations at risk of developing ARDS, but in the absence of a single aetiological agent and given the confusion surrounding the role of different mediator systems in the pathophysiology of an acute lung injury, it has become more difficult to use the label ARDS in a meaningful way. Our own particular approach to this problem is illustrated in Fig. 18.1. The primary

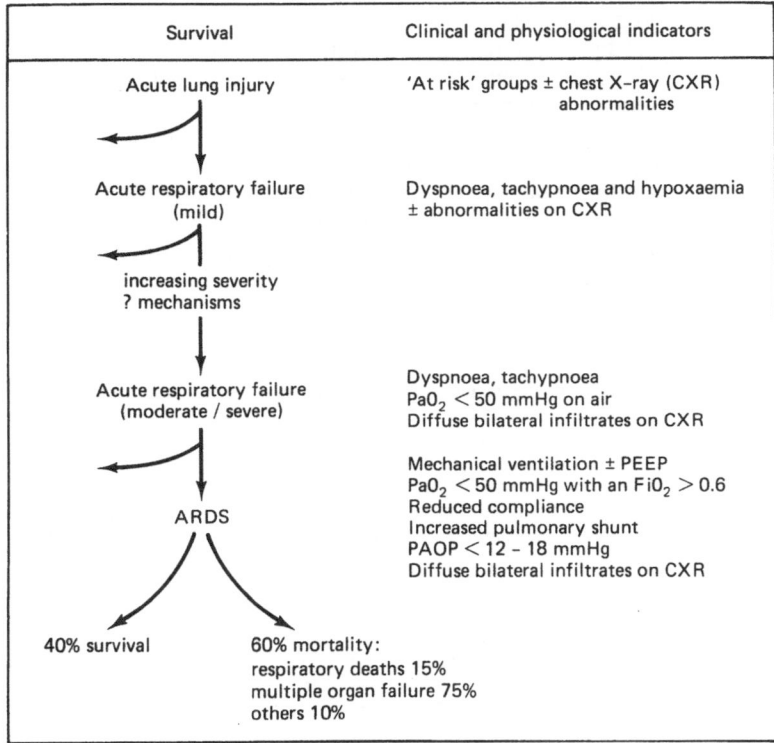

Fig. 18.1. The progression towards an adult respiratory distress syndrome. PaO$_2$, arterial oxygen tension; FiO$_2$, fraction of inspired oxygen; PAOP, pulmonary artery occlusion pressure.

event in the disease process may be best described as an "acute lung injury" which can be associated with a clinical condition that progressively deteriorates through the state of simple mild, moderate and severe "acute respiratory failure" towards full-blown ARDS. Presumably, depending upon the severity of the primary insult, the presence or absence of infection and the form of therapy used in management with the timing of its application, recovery can occur at any stage. Therefore, only with the introduction of standard and accurate methods of scoring the severity of respiratory failure and the disturbance in overall physiology with or without some specific and sensitive assay of some mediator(s) predictive of a severe acute lung injury will these problems of terminology be overcome. Without these, it is impossible to compare the results of the various different clinical trials reported in the literature. At this time it seems to us that it should be impossible to report the survival of a series of patients with ARDS without quoting APACHE II or SAPS scores, defining and documenting the number and severity of organ systems dysfunction and the prevalence of sepsis.

At Risk Groups

As the diagnosis of ARDS has become increasingly common in the ICU, so the number of risk factors and groups of susceptible patients have grown. This may represent not only an increasing awareness of the condition (even perhaps ARDS becoming a "fashionable" diagnosis amongst intensivists-in-training) but also a real increase in prevalence as a greater number of critically ill patients are supported to develop the sequelae of an episode of severe septic shock, massive trauma or surgery, and gastric aspiration. Table 18.1 lists some of the many conditions presently recognized as risk factors. Some, in particular those which directly damage the lung, e.g. the aspiration of gastric contents, are readily recognized and investigations of pulmonary function (serial chest X-rays and arterial blood gas analysis) will be performed as a matter of course. In contrast, patients on the general ward, in whom the acute

Table 18.1. Some predisposing factors for the development of ARDS

1. Direct pulmonary injury	Disseminated intravascular coagulation
Pulmonary infection	Extracorporeal circulations
Pulmonary contusion	Hepatic failure, uraemia
Aspiration of stomach contents	Diabetic ketoacidosis
Near-drowning	Thrombotic thrombocytopenic purpura, leukaemia, AIDS
Irritant gas or chemical inhalation	
Smoke inhalation, burn injury	**3. Embolic phenomena**
Oxygen toxicity	Fat embolism syndrome
Radiation injury	Multiple transfusions
	Amniotic fluid emboli
2. Systemic disease processes	
Severe shock of any cause	**4. Systemic toxic effects**
Sepsis	Drugs especially opiates
Massive trauma and burns	Ethylene glycol, Paraquat
Pancreatitis	Lymphangiography, chemotherapy

lung injury occurs as part of a distant pathophysiological process, may not receive the same attention and disturbances in pulmonary function may initially go unnoticed, and hence untreated. This is of considerable importance if some prophylactic or therapeutic intervention—the administration of steroids, non-steroidal anti-inflammatory drugs, vasodilator prostaglandins, CPAP by facemask etc.—is planned in an attempt to prevent the occurrence of a fully developed ARDS.

A recent prospective study of 993 patients thought to be at risk of developing the syndrome allocated cases into eight groups and determined prevalence and fatality rates (Fowler et al. 1983). Prevalence varied between 1.7/100 patients undergoing cardiopulmonary bypass surgery to 35.6/100 patients after aspiration of gastric contents (Table 18.2). As one might expect, prevalence rates were greater in patients who fell into two or more risk groups (24.6 cases of ARDS/100 patients compared with 5.8 cases/100 patients in those with one risk factor only). Of interest was the observation that there was no difference in fatality rates between patients with single or multiple dispositions, which is consistent with the hypothesis of a common systemic process underlying the development of the syndrome. These findings have been confirmed in principle by a second independent study from the Seattle group, although the overall prevalence of ARDS tended to be higher (Pepe et al. 1982).

Table 18.2. Prevalence rates for ARDS in eight preselected groups of patients (from Fowler et al. 1983)

Clinical factor	Prevalence of ARDS (%)
Cardiopulmonary bypass	1.7
Burns	2.4
Bacteraemia	3.8
Massive transfusion	4.6
Multiple fractures	5.3
Pneumonia in the ICU	11.9
Disseminated intravascular coagulation	22.2
Pulmonary aspiration	35.6
Single risk patients	6
Multiple risk patients	25

PEEP and Oxygen Toxicity

The maintenance of adequate arterial oxygenation so that tissue oxygen delivery is not compromised is central to the management of patients with an acute lung injury. Since Ashbaugh and Petty commented that PEEP was "most helpful in combatting atelectasis and hypoxaemia", its use has become widespread in the management of patients with ARDS. Its value in therapy is undisputed, especially since the introduction of the concept of "best PEEP" for individual patients based upon maximal improvements in pulmonary compliance and the greatest oxygen delivery to tissues (Suter et al. 1975; Weisman et al. 1982). Nevertheless, its role and that of CPAP in prophylaxis remain unproven. Theoretically, the prevention of atelectasis with the increased functional residual capacity and mean airway pressure leading to improved arterial oxygenation thereby enabling a reduction in the concentration of inspired oxygen might contribute to a reduction in "iatrogenic" lung damage. The prevention of hypoxic pulmonary arterial vasoconstriction might also reduce right ventricular work load (although PEEP itself tends to increase pulmonary vascular resistance).

Despite these considerations there has been no well-designed, randomized and prospective study which has documented a reduction in the incidence of ARDS by the prophylactic use of PEEP in patients at risk. Pepe et al. (1984), who prospectively studied 92 ventilated patients with accepted risk factors of ARDS, were unable to demonstrate any reduction in the prevalence of the syndrome in those in whom 8 cm H_2O of PEEP was applied early in the course of their illness. However, the prevalence of ARDS in both treated and control groups of this study was significantly higher than expected from risk factor analysis alone. This suggests that the patients included in the investigation were further down the path from acute lung injury to ARDS than initially supposed and does not necessarily preclude an effect of the earlier application of PEEP (or CPAP in the spontaneously breathing individual) on the disease process. It seems obvious that the selection of patients on the basis of a requirement for endotracheal intubation and mechanical ventilation may introduce a bias for the inclusion of more severe cases of respiratory failure (Pepe 1986).

In the absence of evidence for a protective effect, there have been suggestions that PEEP might actually be detrimental (Albert 1985). Apart from the well-recognized complications of increasing peak and mean airway pressures—direct barotrauma, an increase in dead space ventilation, reductions in cardiac output and oxygen delivery to tissues despite improving arterial oxygenation, a redistribution of blood flow away from vital organs—interest has centred upon the effects of PEEP on reductions in nutrient bronchial artery blood flow and the development of bronchiolectasis (Baile et al. 1984; Navaratnarajah et al. 1984). Various animal studies have suggested that high levels of PEEP may be detrimental to overall pulmonary homeostasis and increase lung interstitial water, but there is little evidence in humans that a moderate level of PEEP aggravates the lung injury or may by itself be a risk factor for ARDS (Weisman et al. 1982; Rounds and Brody 1984).

Against this background remains the incontestable fact that the application of PEEP or CPAP permits a reduction in the inspired concentration of oxygen required to correct the severe arterial hypoxaemia of ARDS. A high concentration of oxygen administered for a prolonged period of time is a potent pulmonary toxin (Deneke and Fanburg 1980) and has been considered a risk factor for the development of the syndrome (Witschi et al. 1981). The experimental data concerning the toxic effects of oxygen are again predominantly based upon various animal models but it is well known that human volunteers breathing 100% oxygen at 1 atmosphere suffer a fall in vital capacity after 24–48 hours. Morphological changes reminiscent of those occurring in ARDS—hyaline membrane deposition, fibrin exudates and interstitial oedema (Pratt et al. 1979)—have been documented in primates breathing 100% oxygen for as little as 12–24 hours. Whether oxygen toxicity is in fact a predominant factor in the propagation of an acute lung injury in humans remains an open question but it seems wise to avoid this particular risk.

Of course, it may be impossible to oxygenate some patients adequately without using high concentrations of oxygen and in this situation 100% oxygen may be essential for the short-term survival of an individual receiving treatment. Nevertheless, in the quest for "best ventilation" in ARDS we prefer to manipulate the mean airway pressure through changes in the inspiratory to expiratory time (I:E) ratio and level of PEEP rather than prolong the administration of a presumed pulmonary toxin. The aim is to maintain effective oxygen delivery to tissues and each patient tends to respond differently to manipulations of these three variables of mechanical ventilation. Certainly, the arterial oxygen tension alone is not the end-point for the assessment of adequate ventilation or any other form of respiratory support and a certain level of hypoxaemia may have to be accepted in order to prevent pulmonary barotrauma or oxygen toxicity to the pulmonary epithelium.

The Diagnosis of ARDS

Since the clinical presentation of patients with ARDS has changed little from its original description in 1967, the development of physiological indices for an accurate definition of the syndrome has been essential for the standardization of "entry criteria" into various clinical studies. These additional features of clinical diagnosis usually include the presence or absence of PEEP/CPAP, specific values for the alveolar–arterial oxygen tension gradient (or the arterial : alveolar ratio) and calculated pulmonary shunt fraction, and measurements of pulmonary compliance. Some of these entry criteria are summarized in Table 18.3.

Similarly, exclusion criteria have been developed so that most studies attempt to ensure that patients with cardiogenic oedema or pre-existing pulmonary disease are not included. Patients with ARDS usually present with complex fluid-balance problems and frequently have received large volumes of crystalloid and colloid as a part of their general resuscitation. Ashbaugh and Petty emphasized that the chest X-ray appearances of their patients "were frequently confused with acute cardiac failure and mild pulmonary oedema" and they were unable to define the role of volume overload in the pathogenesis of the respiratory failure.

Following the introduction of bedside pulmonary artery catheterization with the measurement of pulmonary arterial occlusion pressure (PAOP) as a guide to pulmonary capillary and left atrial pressure, it has been possible to quantify to a certain degree the contribution of raised capillary pressure to the formation of pulmonary oedema. Indeed, it has been thought essential to measure PAOP in ARDS, especially as these patients tend to develop pulmonary hypertension and right ventricular dysfunction limiting the usefulness of measurements of central venous pressure alone. This new technology, although widely applied, has brought with it a number of problems (Robin 1985). Firstly, it is not clear that the PAOP is an accurate measurement of either capillary pressure or left ventricular filling pressure in ARDS, especially in the presence of PEEP (Cope et al. 1986; Sibbald et al. 1986). Secondly, although a PAOP of less than 12–18 mmHg is a frequent "entry" requirement for the diagnosis of ARDS, this is rarely related to the prevailing plasma colloid oncotic pressure—which may fall precipitously in sepsis and in patients undergoing vigorous crystalloid resuscitation (Twigley and Hillman 1985). The vexed question of whether excessive crystalloid administration contributes to the apparently higher incidence of ARDS in some North American centres (where weight gains of 15–20 kg are not unusual in patients resuscitated from shock) remains unanswered. Finally, since pulmonary artery catheterization is now reserved in most units for the most severely ill, a selection bias for the entry of more severe cases of ARDS into clinical studies has occurred.

Table 18.3.   Some diagnostic criteria used to define acute respiratory failure and ARDS

Investigators and study	Criteria used in diagnosis
Ashbaugh et al. (1967) (original description of ARDS)	Dyspnoea, tachypnoea, refractory hypoxaemia, reduced pulmonary compliance and bilateral alveolar infiltration on CXR
Pontoppidan et al. (1985) (MGH scoring system for acute respiratory failure)	Four categories of respiratory failure: (i) *At risk:* one or more risk factors, no/minimal changes on CXR, oxygen by face mask for short periods only (ii) *Mild:* minimal diffuse or lobar changes on CXR with or without intubation and CPAP with an FiO_2 <0.5 (iii) *Moderate:* panlobar alveolar infiltrates of one or both lungs or CXR, intubated for more than 24 hours requiring positive airway pressure (CMV, IMV, PEEP, CPAP) with an FiO_2 >0.5 (iv) *Severe:* bilateral panlobar alveolar infiltrates on CXR, intubated requiring CMV and PEEP, PaO_2 <50 mmHg with FiO_2 of 1.0 for 8 hours or more, or FiO_2 >0.6 for more than 48 hours
Bartlet et al. (1986) (NIH study of acute hypoxic respiratory failure, 1975–1977)	Endotracheal intubation and positive airway pressure for at least 24 hours with an FiO_2 >0.50 Five diagnostic categories defined: pneumonitis, capillary leak syndromes (ARDS), high-pressure oedema, thromboemboli and chronic airways obstruction
Zapol et al. (1979) (extracorporeal membrane oxygenation study)	Fast or slow entry criteria (i) *Fast entry:* PaO_2 <50 mmHg for more than 2 hours with an FiO_2 of 1.0 and PEEP >5 cm H_2O (ii) *Slow entry:* despite maximal medical therapy for 48 hours, PaO_2 <50 mmHg for more than 12 hours when measured on an FiO_2 >0.6 and PEEP >5 cm H_2O with a pulmonary shunt >30% measured on PEEP with an FiO_2 of 1.0
Holcroft et al. (1986) (prostaglandin E_1 study)	Patients with ARDS requiring mechanical ventilation with an FiO_2 >0.4 and PEEP >5 cm H_2O "not responding to conventional therapy"
The Upjohn European multicentre trial of prostaglandin E_1 in ARDS (in progress)	Patients with postoperative ARDS or associated with sepsis or traumá requiring mechanical ventilation, with: an FiO_2 >0.5 ± PEEP in two or more determinations at least 2 hours apart in the preceding 24 hours giving a PaO_2 <75 mmHg or an arterial : alveolar ratio <0.3 with a pulmonary arterial occlusion pressure <18 mmHg
European Society of Intensive Care study of acute respiratory failure (in progress)	Patients with respiratory distress and a CXR of diffuse bilateral pulmonary infiltrates and severe hypoxaemia defined by: PaO_2 <80 mmHg with an FiO_2 >0.6 for at least 24 hours

CXR, chest X-ray; PaO_2, arterial oxygen tension; FiO_2, inspired fraction of oxygen; CMV, continuous mandatory ventilation; IMV, intermittent mandatory ventilation.

Rinaldo has recently emphasized this last point in so far as many more patients with acute respiratory failure in the Pittsburgh ICU are managed clinically without the recourse to pulmonary artery catheterization. Although such patients may fulfil the broad diagnostic criteria for ARDS, they are ineligible for entry into any study requiring the measurement of PAOP. Yet these patients (which include all of the trauma-associated ARDS group) have a markedly increased survival rate of 70% (Rinaldo 1986). Again, this sort of observation makes comments concerning the unchanging mortality from ARDS meaningless. It seems that two quite distinct sets of diagnostic criteria are in operation: the first, a broad clinical category including many mild or moderate cases of acute respiratory failure in association with one or more risk factors for ARDS; the second depending upon more sharply defined physiological criteria, selecting patients with severe acute respiratory failure for single or multi-centre "publishable" studies of ARDS. Sadly, this distinction may lead to erroneous conclusions concerning the therapeutic efficacy of various forms of respiratory support and drugs, i.e. new therapeutic regimens will only be tested in groups of patients with a predetermined very high mortality rate. Any effective regimen will have to be *highly* effective if it is to reduce the mortality rate in this form of acute respiratory failure.

Conclusion

In the absence of specific serological markers, the diagnosis of ARDS remains essentially a form of pattern recognition. The pathogenesis of acute lung injury associated with one or more risk factors for ARDS has once again opened up since the pre-eminent role of complement activation and the sequestration of leucocytes within the lung has been called into question (Braude et al. 1985; Ognibene et al. 1986; Rinaldo and Rogers 1986). In the absence of a unifying hypothesis of pathogenesis, the labelling of a clinical situation as "ARDS" is at best an attempt to communicate the severity of an acute lung injury to other investigators in the field, but at worst is misleading, suggesting a well-understood, distinct disease process primarily affecting the lungs. This is patently not the case, and advances in therapy now depend upon the identification of some specific mechanism(s) of the widespread tissue injury so that steps to inhibit its activation or block or reverse its effects can be taken early in the course of respiratory failure.

References

Albert R (1985) Least PEEP: *primum non nocere*. Chest 87:2–4
Andreadis N, Petty T (1985) Adult respiratory distress syndrome: problems and progress. Am Rev Respir Dis 132:1344–1346
Anonymous (1986a) Adult respiratory distress syndrome. Lancet I:301–303
Anonymous (1986b) ARDS: a clinical view. Lancet II:439
Ashbaugh D, Bigelow D, Petty T, Levine B (1967) Acute respiratory distress in adults. Lancet I:319–323
Baile E, Albert R, Kirk W, Lakshaminarayan S, Wiggs B, Pare P (1984) Positive end expiratory pressure decreases bronchial blood flow in the dog. J Appl Physiol 56:1289–1293
Bartlet R, Morris A, Fairley H, Hirsch R, O'Connor N, Pontoppidan H (1986) A prospective study of acute hypoxic respiratory failure. Chest 89:684–689
Bell R, Coalson J, Smith J, Johanson W (1983) Multiple organ system failure and infection in adult respiratory distress syndrome. Ann Intern Med 99:293–298
Braude S, Apperley A, Krausz T, Goldman J, Royston D (1985) Adult respiratory distress syndrome after allogeneic bone marrow transplantation: evidence for a neutrophil independent mechanism. Lancet I:1239–1242

Cameron S (1986) Acute renal failure in the intensive care unit today. Intensive Care Med 12:64–70

Cope D, Allison R, Parmentier J, Miller J, Taylor A (1986) Measurement of effective pulmonary capillary pressure using the pressure profile after pulmonary artery occlusion. Crit Care Med 14:16–22

Deneke S, Fanburg B (1980) Normobaric oxygen toxicity of the lung. N Engl J Med 303:76–86

Fowler A, Hamman R, Good J, Benson K, Baird M, Eberle D, Petty T, Hyers T (1983) Adult respiratory distress syndrome: risk with common predispositions. Ann Intern Med 98:593–597

Fowler A, Hamman R, Zerbe G, Benson K, Hyers T (1985) Adult respiratory distress syndrome: prognosis after onset. Am Rev Respir Dis 132:472–478

Holcroft J, Vassar M, Weber C (1986) Prostaglandin E1 and survival in patients with adult respiratory distress syndrome. Ann Surg 203:371–378

Hyers T, Fowler A (1986) Adult respiratory distress syndrome: causes, morbidity and mortality. Fed Proc 45:25–29

Montgomery A, Stager M, Carrico C, Hudson L (1985) Causes of mortality in patients with the adult respiratory distress syndrome. Am Rev Respir Dis 132:485–489

Moon W (1948) The pathology of secondary shock. Am J Pathol 24:235–374

Navaratnarajah M, Nunn J, Lyons D, Milledge J (1984) Bronchiolectasis caused by positive end expiratory pressure. Crit Care Med 12:1036–1038

Newman J (1983) ARDS: new insights and unsolved problems. Intensive Care Med 9:303–306

Ognibene F, Martin S, Parker M, Schlesinger T, Roach P, Burch C, Shelhamer J, Parrillo J (1986) Adult respiratory distress syndrome in patients with severe neutropenia. N Engl J Med 315:547–551

Pepe P (1986) The clinical entity of the adult respiratory distress syndrome. Definition, prediction and prognosis. In: Wiedeman H, Matthay M, Matthay R (eds) Acute lung injury. Crit Care Clin 2 (July):377–403

Pepe P, Potkin R, Reus D, Hudson L, Carrico C (1982) Clinical predictors of the adult respiratory distress syndrome. Am J Surg 144:124–130

Pepe P, Hudson L, Carrico C (1984) Early application of positive end expiratory pressure in patients at risk for the adult respiratory distress syndrome. N Engl J Med 311:281–286

Petty T (1985) Indicators of risk, course and prognosis in the adult respiratory distress syndrome. Am Rev Respir Dis 132:471

Pontoppidan H, Geffin B, Lowenstein E (1972) Acute respiratory failure in the adult. N Engl J Med 287:690–698, 743–752, 799–806

Pontoppidan H, Huttemeier P, Quinn D (1985) Etiology, demography and outcome. In: Zapol W, Falke K (eds) Acute respiratory failure. Marcel Dekker, New York, pp 1–21

Pratt P, Vollmer R, Shelburne J, Crapo J (1979) Pulmonary morphology in a multihospital collaborative extracorporeal membrane oxygenation project. Am J Pathol 95:191–214

Rinaldo J (1986) Indicators of risk, course and prognosis in adult respiratory distress syndrome. Am Rev Respir Dis 133:343

Rinaldo J, Rogers R (1986) Adult respiratory distress syndrome. N Engl J Med 315:578–580

Robin E (1985) The cult of the Swan–Ganz catheter: overuse and abuse of pulmonary flow catheters. Ann Intern Med 103:445–449

Ronin E, Carey L, Grenvik A, Glauser F, Gaudio R (1972) Capillary leak syndrome with pulmonary edema. Arch Intern Med 130:66–71

Rounds S, Brody J (1984) Putting PEEP in perspective. N Engl J Med 311:323–325

Seidenfeld J, Pohl D, Bell R, Harris G, Johanson W (1986) Incidence, site and outcome of infections in patients with the adult respiratory distress syndrome. Am Rev Respir Med 134:12–16

Sibbald W, Driedger A, Cunningham D, Cheung H (1986) Right and left ventricular performance in acute hypoxaemic respiratory failure. Crit Care Med 14:852–857

Suter P, Fairley H, Isenberg M (1975) Optimum end-expiratory pressure in patients with acute respiratory failure. N Engl J Med 292:284–289

Twigley A, Hillman K (1985) The end of the crystalloid era? Anaesthesia 40:860–871

Wardle E (1982) Acute renal failure in the 1980s: the importance of septic shock and of endotoxaemia. Nephron 30:193–200

Weisman I, Rinaldo J, Rogers R (1982) Positive end-expiratory pressure in adult respiratory failure. N Engl J Med 307:1381–1384

Wheeler D, Feehally J, Walls J (1986) High risk acute renal failure. Q J Med 61:977–984

Witschi H, Haschek W, Klein-Szanto A, Hakkinen P (1981) Potentiation of diffuse lung damage by oxygen determining variables. Am Rev Respir Dis 123:98–103

Zapol W, Snider M, Hill J, Fallat R, Bartlett R, Edmunds L, Morris A, Pierce E, Thomas A, Proctor H, Drinker P, Pratt P, Bagniewski A, Miller R (1979) Extracorporeal membrane oxygenation in severe acute respiratory failure. JAMA 242:2193–2196

Subject Index